National Aeronautics and Space Administration

NASA's
SPACE LAUNCH
SYSTEM
REFERENCE GUIDE

www.nasa.gov   #Artemis   hsa llc

Version 2

TABLE OF CONTENTS

Table of Contents

From the SLS Program Manager and Chief Engineer 5

Introduction 6

SLS Quick Facts 7

    NASA's Block 1 Space Launch System by the Numbers 8

        SLS Vehicle 8

        Core Stage 8

        RS-25 Engines 8

        Solid Rocket Boosters 9

        Interim Cryogenic Propulsion Stage (Upper Stage) 9

        Launch Vehicle Stage Adapter 9

        Orion Stage Adapter 9

        Secondary payloads 10

    SLS Overview 12

    SLS Design: Chosen from Thousands of Options 13

    Common Elements and Evolvability 15

        More Powerful Rockets for Future Missions 17

    SLS Vehicle-Level Testing 19

    Super Heavy Lifting on the Ground: SLS Transportation, Logistics, and Pathfinders 22

    Exterior Features of SLS 25

        Thermal Protection System 25

        Livery and Photogrammetric Markings 26

        Instrumentation and Vehicle Telemetry 27

The Role of SLS in Launching Artemis I 28

    Artemis I Secondary Payloads 30

The Elements of SLS 34

    Core Stage 35

        Manufacturing, Test, and Checkout 39

        Core Stage Fun Facts 41

CONTINUED TABLE OF CONTENTS

- RS-25 Engines 42
  - Manufacturing, Test, and Checkout 44
  - Flight History of Artemis I RS-25 Engines 45
  - RS-25 Fun Facts 46
- Solid Rocket Boosters 47
  - Manufacturing, Test, and Checkout 50
  - Flight History of the SLS Booster Structures 52
  - Solid Rocket Boosters Fun Facts 53
- Integrated Spacecraft/Payload Element 54
  - Interim Cryogenic Propulsion Stage 55
  - Launch Vehicle Stage Adapter 56
  - Orion Stage Adapter 57
  - Manufacturing, Test, and Checkout 58
- SLS Avionics and Software 59

Management Roles and Facilities 60
- Marshall Space Flight Center 61
  - Unique Test Facilities 61
  - Supporting Launch Operations 62
- Michoud Assembly Facility 64
- Stennis Space Center 66

Industry Partners 68
- Aerojet Rocketdyne 69
- Boeing 70
- Northrop Grumman 71
- Teledyne Brown Engineering 72
- United Launch Alliance 73

Additional Resources 74
- Understanding SLS: Infographics 75
- Acronym List 83

hsa llc

From the SLS Program Manager and Chief Engineer

As Program Manager and Chief Engineer for NASA's Space Launch System (SLS), it has been our honor to shepherd America's new super heavy-lift rocket – a launch vehicle like no other – from design, through testing, manufacturing, assembly, and now launch of the Artemis I test flight. We'd like to thank our team members, stakeholders, and industry partners in this endeavor. This is America's rocket, and we could not have made it to the launch pad without you.

After the Apollo Program ended, NASA focused on learning to live and work in low-Earth orbit – about 250 miles up – with the space shuttle and the International Space Station (ISS). There have been astronauts living continuously onboard the ISS since 2000, and our knowledge of how to keep crew safe, conduct science, and maintain living and laboratory facilities in space has advanced tremendously. Now, NASA's human spaceflight program is once again looking to deep space...to the Moon and then to Mars.

And for that, a new rocket, new spacecraft for crew, and new ground and launch capabilities are needed.

Meet the Space Launch System rocket, the Orion crew spacecraft, and Exploration Ground Systems (EGS) at NASA's Kennedy Space Center in Florida. These technologies are making a new generation of missions to the Moon – the Artemis Generation – possible. Along with the Gateway and the Human Landing System, NASA is preparing to send more astronauts, conduct more science, and explore more of the Moon than ever before.

SLS is the rocket that will launch Orion and its crew to lunar orbit. While the initial configuration, the Block 1 rocket for Artemis I, will launch with 8.8 million lbs. of thrust (more than even the Saturn V), future variants will be even more powerful. Sending astronauts to the Moon 239,000 miles away – one thousand times farther than the space station in low-Earth orbit – is an inherently risky and dangerous undertaking.

We are proud of the extensive testing, from component to entire vehicle, that has gone into making SLS the safest, most reliable rocket for human missions that NASA has ever built. And the Artemis I test flight gives us a chance to test out all the systems in deep space, as well as our procedures on the ground, before we take the next giant leap and once again send astronauts to the Moon.

So, why build SLS? So the world can see a new generation of bootprints in lunar regolith. To advance from a 20th-century Moon shot to a 21st-century Moon stay. To build, with our partners, not just a rocket, but a platform for human missions to the Moon and science missions to explore the solar system, like those of us staring at the stars and heavens have always dreamed of. Why build SLS? To find out not only what's *out there* but what's *in us*.

John Honeycutt
SLS Program Manager

Dr. John Blevins
SLS Chief Engineer

INTRODUCTION

INTRODUCTION

This document was prepared by NASA's SLS Program Office at NASA's Marshall Space Flight Center, in Huntsville, Alabama, which has responsibility for design, development, testing, and engineering of SLS, the new super heavy-lift rocket that will send astronauts to the Moon as part of the Artemis missions. The first flight of SLS and the Orion crew spacecraft, Artemis I, will lift off from NASA's Kennedy Space Center in Florida and send an uncrewed Orion to lunar orbit. Artemis I is a rigorous test flight, designed to thoroughly test all systems of the SLS rocket prior to crewed flights commencing with the Artemis II mission.

This document is designed to serve as a general reference for the SLS rocket's initial Block 1 configuration used for NASA's early Artemis missions. SLS capabilities will be better understood and refined as the program analyzes early flight data and the SLS configuration evolves. In addition to providing basic facts about the SLS Block 1 rocket configured to send the Orion spacecraft to the Moon, this document also includes information on SLS design, capabilities, major components, and activities such as manufacture, test, assembly, launch, and mission.

Dimensional and performance data used throughout this document are approximate for mission-specific, security, and proprietary reasons. Dimensions and measures are provided in United States customary units as well as metric units in parenthesis.

In the Table of Contents, click on the topic to jump to that section. For quick navigation in the document, click on the gray up arrows at the bottom of each page to return to the Table of Contents.

SLS QUICK FACTS

SLS represents a bold new vision for NASA's human spaceflight program. In order to make a new generation – the Artemis Generation – of crewed missions to the Moon possible, the SLS rocket uses proven propulsion systems consisting of solid rocket boosters and liquid-fuel RS-25 engines mated to a new central core stage.

SLS uses larger solid rocket boosters than the space shuttles and liquid hydrogen/liquid oxygen-fed RS-25 engines opera-ting at a higher thrust level and with new controllers.

The core stage, an all-new development consisting of propellant tanks, avionics, and related equipment, houses the four RS-25s and provides attach points for the boosters. Above the core stage, the Interim Cryogenic Propulsion Stage (ICPS) provides in-space propulsion. The launch vehicle stage adapter partially encloses the ICPS and changes the diameter of the rocket. The Orion stage adapter, located between SLS and the Orion crew vehicle, contains CubeSat payloads for the Artemis I mission and connects the rocket to the Orion spacecraft.

The first section of this guide consists of quick facts and introductory information. The following sections go into more detail about each of the elements of the Block 1 SLS and include additional illustrations, photos, tables, and additional infographics.

The images and illustrations in this guide are all NASA images. For downloadable versions of these and many other SLS images, go to: https://www.nasa.gov/exploration/systems/sls/multimedia/images.html

Elements of SLS and Orion

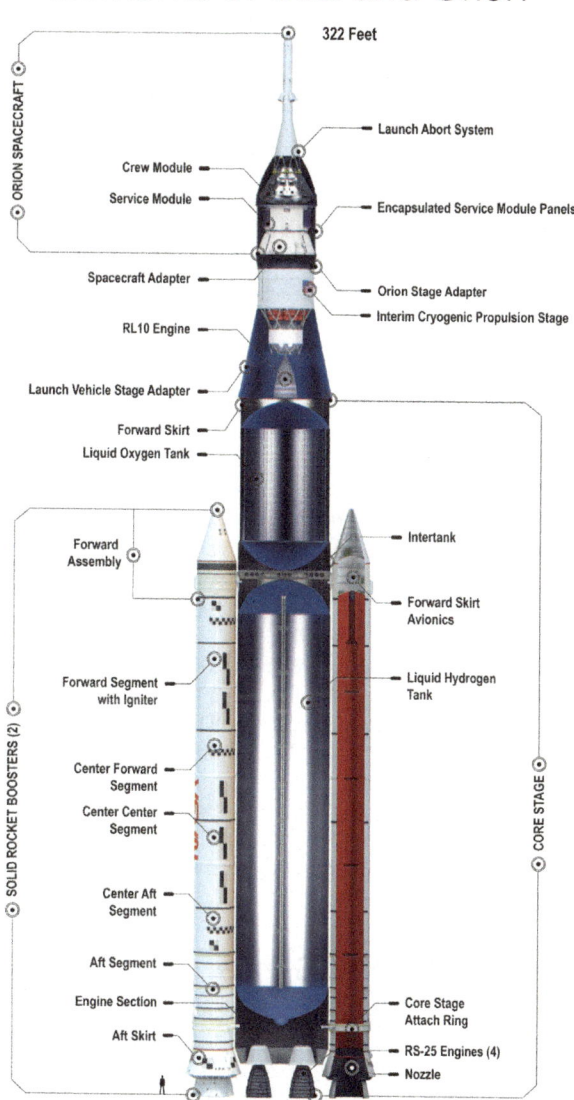

This cutaway illustration identifies the major elements of the SLS and Orion stack. The SLS portion consists of, from the ground up: twin solid rocket boosters with major components labeled, RS-25 liquid-fuel engines, central core stage with propellant tanks and flight computers, launch vehicle stage adapter, ICPS, and Orion stage adapter. Orion is atop SLS. A 6-ft. figure is shown for comparison in the lower left.

SLS QUICK FACTS
NASA'S SLS BLOCK 1 BY THE NUMBERS

SLS Block 1 by the Numbers*

SLS Vehicle

Vehicle design: Evolvable super heavy-lift
Height: 322.4 ft. (98.3 m)
Weight: 5.74 million lbs. (2,603 t) fueled
3.5 million lbm (1,588 t) unfueled
Main propulsion: Four RS-25 liquid propellant engines and two five-segment solid rocket boosters
Maximum thrust: 8.8 million lbs. (39,144 kN)
Launch thrust: 8.27 million lbs. (36,786 kN)
Maximum speed: 22,670 mph (36,484 km/h) at ICPS trans-lunar injection TLI) main engine cutoff (MECO)
Single-launch payload to low-Earth orbit: 209,439 lbs. (95 t)
Payload to TLI: > 59,525 lbs. (27 t)

Core Stage

Contractor: Boeing
Height: 212 ft. (64.6 m) from forward skirt to engine exhaust exit plane
Diameter: 27.6 ft. (8.4 m)
Weight (without engines): 2.4 million lbs. (1,088 t) fueled, 188,000 lbs. (85.3 t) unfueled
Capacities: 537,000 gal. (2 million L),
317,000 lbs. (143.8 t) liquid hydrogen fuel;
196,000 gal. (741,941 L),
1.86 million lbs. (843.7 t)
liquid oxygen oxidizer
Maximum thrust: Approximately 2 million lbs. (8,896 kN)
Burn Time: 480 sec.

RS-25 Engines

Contractor: Aerojet Rocketdyne
Height: 14 ft. (4.3 m)
Diameter: 8 ft. (2.4 m)
Weight (each): 7,750 lbs. (3.5 t)
Propellants: Liquid hydrogen, liquid oxygen
Thrust: 418,000 lbs. (1,852 kN) at launch;
512,300 lbs. (2,279 kN), maximum at 109 percent power level
Burn time: 480 sec.

Stacked Artemis I rocket, consisting of SLS and Orion, on the mobile launcher.

The Artemis I core stage waits in the Vehicle Assembly Building transfer aisle before stacking on the mobile launcher.

The Artemis I RS-25 engines installed in the core stage during assembly at NASA's Michoud Assembly Facility in New Orleans.

*Data are Approximate

SLS QUICK FACTS
NASA'S SLS BLOCK 1 BY THE NUMBERS

Solid Rocket Boosters

Contractor: Northrop Grumman
Height: 177 ft. (54 m)
Diameter: 12 ft. (3.7 m)
Weight (each): 1.6 million lbs. (726 t) loaded, 219,000 lbs. (99.3 t) empty
Solid rocket motor: Five propellant segments
Propellant: Polybutadiene acrylonitrile (PBAN)
Thrust: 3.3 million lbs. (14,679 kN) each at launch; 3.6 million lbs. (16,014 kN) each maximum
Burn time: 126 sec.

Artemis I five-segment solid rocket boosters stacked on the mobile launcher.

Interim Cryogenic Propulsion Stage (Upper Stage)

Contractor: Boeing/United Launch Alliance
Designation: Interim Cryogenic Propulsion Stage (modified Delta Cryogenic Second Stage)
Height: 45 ft. (13.7 m)
Diameter: 16.7 ft. (5.1 m)
Weight: 72,197 lbs. (32.7 t) fueled; 8,200 lbs (3.7 t) unfueled
Engine: Aerojet Rocketdyne RL10B-2, RL10C-2 (Artemis II/III)
Propellants: Liquid hydrogen/liquid oxygen
Maximum thrust: 24,750 lbs. (110 kN)
Reaction Control System (RCS): Hydrazine

The Artemis I ICPS at Kennedy prior to stacking.

Launch Vehicle Stage Adapter

Contractor: Teledyne Brown Engineering
Height: 27.5 ft. (8.4 m)
Diameter: 27.5 ft. (8.4 m) bottom, 16.5 ft. (5.0 m) top
Weight: 10,000 lbs. (4.5 t)

The Artemis I launch vehicle stage adapter arrives at the Vehicle Assembly Building for stacking on the Artemis I rocket.

Orion Stage Adapter

Contractor: NASA Marshall Space Flight Center
Height: 5 ft. (1.5 m)
Diameter: 18 ft. (5.5 m)
Weight: 1,800 lbs. (0.8 t)
Available volume for payloads: 516 ft.^{3} (14.6 m^3)

Orion stage adapter complete and ready for CubeSats.

*Data are Approximate

SLS QUICK FACTS
SECONDARY PAYLOADS

Secondary Payloads

Ten CubeSat payloads for Artemis I, each measuring 4.4 in. (11.1 cm) x 9.4 in. (22.8 cm) x 14.4 in. (36.6 cm); each CubeSat is 6U (1U [unit] = 10 cm x 10 cm x 10 cm), not to exceed 30 lbs. (14 kg)

Moon

Lunar IceCube – Morehead State University, Kentucky
Searching for water in all forms and other volatiles with an infrared spectrometer

LunaH-Map – Arizona State University, Arizona
Creating higher-fidelity maps of near-surface hydrogen in craters and other permanently shadowed regions of the lunar South Pole with neutron spectrometers

OMOTENASHI – Japanese Aerospace Exploration Agency (JAXA), Japan Developing world's smallest lunar lander and studying lunar environment

LunIR – Lockheed Martin, Colorado
Performing advanced infrared imaging of the lunar surface

Sun

CuSP – Southwest Research Institute, Texas
Measuring particles and magnetic fields as a space weather station

Asteroid

NEA Scout – NASA's Marshall Space Flight Center, Alabama
Traveling by solar sail to a near-Earth asteroid and taking pictures and other characterizations of its surface

Earth

EQUULEUS – University of Tokyo/JAXA, Japan
Imaging Earth's plasmasphere for a better understanding of Earth's radiation environment from Earth-Moon Lagrange 2 point OR Lagrange Point 2

The Artemis I Orion stage adapter with mounting brackets for secondary payloads.

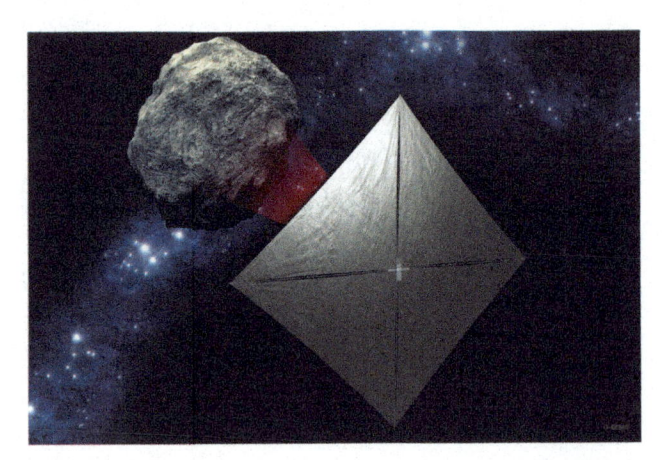

Artist concept of Artemis I secondary payload, NEA Scout, with its deployed solar sail on its way to an asteroid.

Preparing the NEA Scout payload for its flight aboard Artemis I.

SLS QUICK FACTS
SECONDARY PAYLOADS

Other Missions

BioSentinel – NASA's Ames Research Center, California
Using single-celled yeast to detect, measure, and compare the impact of deep space radiation on living organisms over a long period of time

ArgoMoon – European Space Agency (ESA)/ASI, ArgoTec, Italy
Observing the ICPS with advanced optics and software imaging system

Centennial Challenges

Team Miles – Florida
Demonstrating propulsion using plasma thrusters and competing in NASA's Deep Space Derby

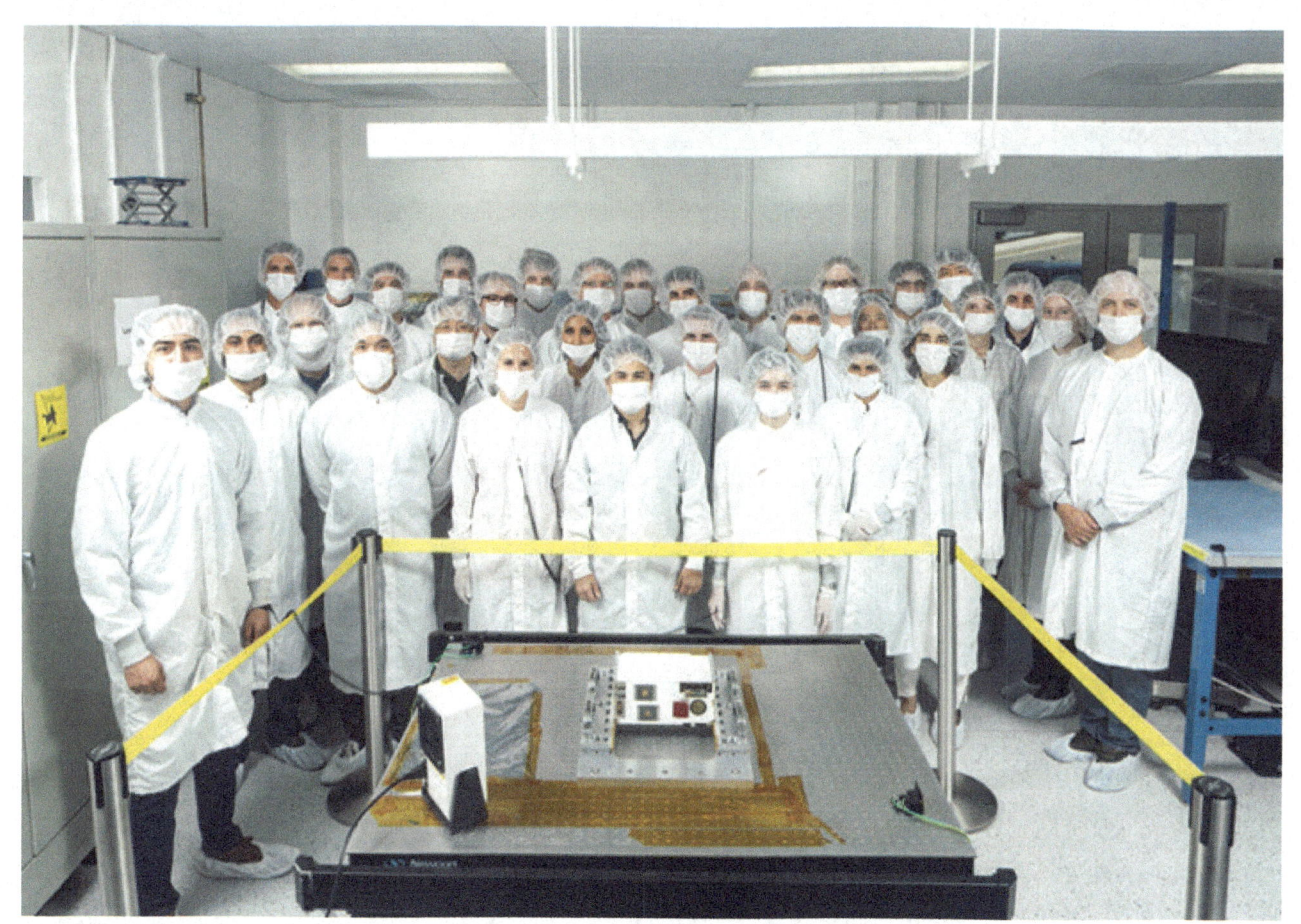

The NEA Scout team at NASA's Jet Propulsion Laboratory, which supplied science instruments to characterize a target near-Earth asteroid.

SLS OVERVIEW

NASA's SLS is the world's most powerful rocket and the backbone of NASA's human lunar exploration program. No other rocket can send astronauts and the Orion spacecraft directly to the Moon for the Artemis missions. SLS provides an unmatched capability to deliver greater mass and volume than any current launch vehicle for both human and robotic exploration of the Moon, Mars, and the outer planets.

SLS was established by the NASA Authorization Act of 2010. The program was created at Marshall in 2011 and received funding in FY2012. SLS is the world's first exploration-class launch vehicle since the Apollo Program's Saturn V.

Along with SLS, NASA's Exploration Systems Development Mission Directorate in the Human Exploration and Operations Mission Directorate is developing the EGS and the Orion spacecraft for crew.

Orion, managed at NASA's Johnson Space Center in Houston, is a spacecraft designed to carry astronauts on exploration missions into deep space. EGS has converted facilities at Kennedy into a next-generation spaceport capable of supporting launches by multiple types of vehicles.

SLS is designed to evolve from an initial Block 1 configuration for the early Artemis uncrewed test and crew missions to a Block 1B vehicle capable of dual crew/cargo missions or dedicated large cargo missions. The ultimate Block 2 configuration design can provide the same dual crew/cargo and cargo-only configurations with even greater payload mass and volume capability.

SLS OVERVIEW
SLS DESIGN

SLS Design: Chosen From Thousands of Options

Super heavy-lift has been validated by numerous NASA and external studies, such as the Space Transportation Architecture Study in 1986 and the Exploration Systems Architecture Study in 2005. As NASA looked toward the space shuttle's retirement, the agency focused on a return to human deep space exploration and the needed super heavy-lift capability to make those missions a reality.

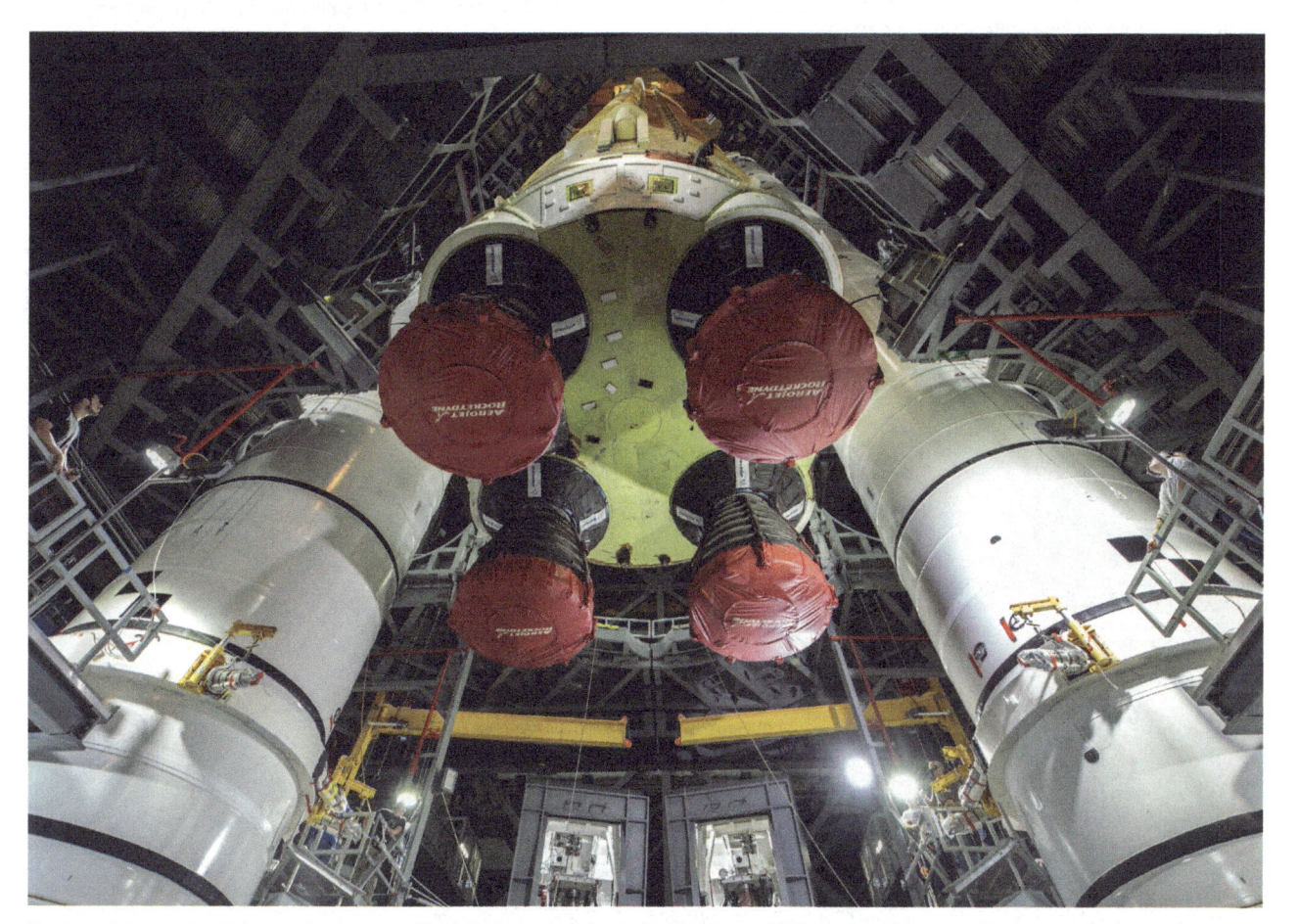

The Artemis I core stage is lowered between the SLS solid rocket boosters in the Vehicle Assembly Building. The SLS design uses a proven propulsion system of solid rocket boosters and liquid-fuel RS-25 engines to launch more payload to the Moon than any rocket since the Saturn V.

SLS OVERVIEW
SLS DESIGN

In defining the vehicle that would become SLS, NASA evaluated thousands of combinations of attributes such as propulsion systems; stages; boosters; performance; development and operations cost; mission complexity, reliability, and risks; and the ability to maintain critical industry base skills. The result: SLS. The evolvable rocket, which can accommodate crew or cargo (with payload fairing) configurations, with a proven propulsion system, provides a safe, human-rated launch capability.

SLS benefits from NASA's half-century of experience with efficient liquid oxygen and liquid hydrogen propellants and from advances in technology and manufacturing practices. Further, by using common design elements, the SLS interfaces with the ground systems at Kennedy, with the Orion spacecraft, and with future payloads will remain consistent over time, reducing complexity of deep space missions.

The SLS operational scheme takes advantage of resources established for the space shuttle, including the workforce, manufacturing processes, tooling and facilities, transportation logistics, launch infrastructure, and liquid oxygen/liquid hydrogen propellants. By using heritage shuttle engines and boosters upgraded for SLS, the program saved the time and cost typical in developing new propulsion systems, and the SLS Program begins flights with proven engine and booster hardware transferred from the Space Shuttle Program.

RS-25 hot fire testing at NASA's Stennis Space Center in Bay St. Louis, Mississippi, helped qualify engines to extreme SLS environments.

With the largest payload mass and volume of any existing rocket, SLS represents greater safety, less risk, and increased probability of mission success in the dynamic, unforgiving environments of spaceflight. More mass and volume translate into fewer launches of fewer pieces requiring less assembly time in orbit.

On the ground, it also translates into simpler hardware design and less transportation, storage, processing, launch operations, and launch pad turnaround activities. SLS was designed for the most challenging deep space missions involving strategic commitments of national resources, national prestige, and human life. SLS represents the best balance of mission performance, safety, cost, affordability, and risk.

Common Elements and Evolvability

SLS evolvability supports a robust resumption of human exploration beyond low-Earth orbit and an expeditious growth path as exploration missions become more complex and demanding.

All variants of the basic SLS design are based on common propulsion components:

- A central core stage housing propellant tanks, engines, avionics, and attach points for a pair of solid rocket boosters.

- Four liquid propellant engines powered by cryogenic liquid hydrogen and cryogenic liquid oxygen from the core stage.

- Two solid-fuel, side-mounted booster rockets that provide the majority of thrust and steering for the rocket during the first two minutes of flight, after which they are jettisoned.

- An in-space upper stage fueled by liquid hydrogen and liquid oxygen.

SLS Evolutionary Path

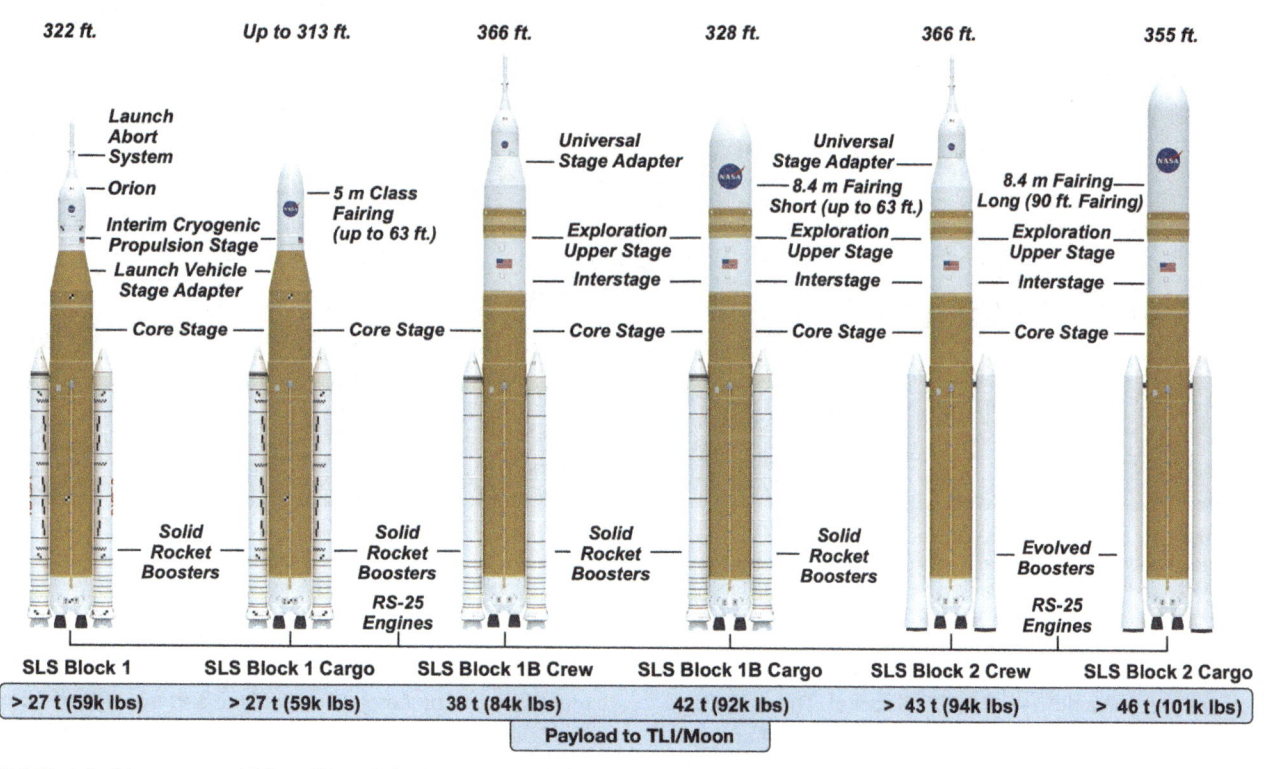

	SLS Block 1	SLS Block 1 Cargo	SLS Block 1B Crew	SLS Block 1B Cargo	SLS Block 2 Crew	SLS Block 2 Cargo
Height	322 ft.	Up to 313 ft.	366 ft.	328 ft.	366 ft.	355 ft.
Payload to TLI/Moon	> 27 t (59k lbs)	> 27 t (59k lbs)	38 t (84k lbs)	42 t (92k lbs)	> 43 t (94k lbs)	> 46 t (101k lbs)

SLS Block 1 with an uncrewed Orion will launch the Artemis I mission. Following the first three Artemis missions, SLS will evolve to a more powerful Block 1B configuration with a new Exploration Upper Stage replacing the ICPS. The ultimate variant, Block 2, will incorporate evolved boosters to increase payload mass to TLI to at least 101,000 lbs. (46 t).

SLS OVERVIEW
COMMON ELEMENTS AND EVOLVABILITY

Each succeeding SLS block variant grows more capable through upgrades to the engines, boosters, and upper stage, providing a flexible platform for a variety of human and robotic deep space missions, rather than requiring development of entirely new rockets to increase performance. In addition to flying Orion, SLS can also be outfitted with wide-diameter payload fairings in varying lengths.

The uncrewed Artemis I mission will be launched on an SLS Block 1 vehicle. Artemis II and Artemis III, the first crewed SLS/Orion missions, also will be launched on Block 1 vehicles. The Block 1 SLS rocket towers 322.4 ft. (98.3 m), and weighs 5.74 million lbs. (2,603 t) fueled. During ascent, the rocket generates 8.8 million lbs. (39,144 kN) of thrust – more than the Saturn V that launched the Apollo missions to the Moon. SLS will transport approximately 59,525 lbs. (27 t) to trans-lunar injection.

The 212-ft. (64.6 m) core stage for Artemis I in the transfer aisle of the Vehicle Assembly Building in preparation for lifting and integration on the mobile launcher. Manufactured by Boeing at Michoud, the stage houses the four RS-25 engines, propellant tanks, and flight computers. Too large to ship over land, NASA's barge Pegasus transported the Artemis I core stage to Stennis for a 'Green Run' test campaign and then to Kennedy for integration with the rest of the vehicle.

The core stage, manufactured by Boeing at Michoud, serves as the backbone of the SLS rocket. The core stage propellant tanks hold 537,000 gal. (2 million L) of cryogenic liquid hydrogen and 196,000 gal. (741,941 L) of cryogenic liquid oxygen. Total propellant weight is 2.2 million lbs. (998 t). The stage will operate for approximately 480 seconds, after which separation from the ICPS, Orion stage adapter, and Orion will occur.

The core stage engine section contains four RS-25 engines manufactured by Aerojet Rocketdyne and transferred from the Space Shuttle Program with new controllers and minor modifications for SLS requirements.

**SLS OVERVIEW
COMMON ELEMENTS AND EVOLVABILITY**

The solid rocket boosters, manufactured by Northrop Grumman, are five-segment solid propellant boosters based on the space shuttle's four-segment boosters. The SLS boosters also use hardware transferred from the Space Shuttle Program, including steel cases, nose cones, frustums, aft skirts, and more. The SLS solid rocket boosters are the largest and most powerful solid rocket boosters ever built for spaceflight. In addition to the extra propellant segment, the SLS boosters incorporate new insulation, avionics, modified propellant grain design, and nozzle design.

The upper stage on the SLS Block 1 is the ICPS, a commercial stage modified for the SLS mission. The ICPS is a liquid hydrogen/liquid oxygen system with a single Aerojet Rocketdyne RL10 engine. The ICPS, manufactured by Boeing/United Launch Alliance (ULA) in Decatur, Alabama, performs the trans-lunar injection engine burn that sends Orion on its course to the Moon. It also contains the avionics to fly the mission after core stage separation until Orion separates from the ICPS.

The ICPS is partially enclosed by a launch vehicle stage adapter, which has a separation system that fires to separate itself and the core stage from the ICPS, Orion stage adapter, and Orion. On top of the ICPS, the Orion stage adapter connects the Orion spacecraft to SLS. The Orion stage adapter will also carry the Artemis I CubeSat payloads to deep space. These small science experiments and technology demonstrations will be deployed after Orion separates from the ICPS and Orion stage adapter.

> **SLS: SUPER HEAVY-LIFT ROCKET FOR GAME-CHANGING SPACE EXPLORATION MISSIONS**
>
> – Block 1 in its crew configuration (with Orion) will transport more than 209,000 lbs. (95 t) to low-Earth orbit and more than 59,000 lbs. (27 t) to TLI.
>
> – Block 1B's crew variant will launch more than 231,000 lbs. (105 t) to low-Earth orbit and more than 83,000 lbs. (38 t) to TLI.
>
> – Block 2 in the crew configuration will carry more than 286,000 lbs. (130 t) to low-Earth orbit and more than 94,000 lbs. (43 t) to TLI.
>
> – The design for each variant's cargo version is capable of launching thousands more pounds to low-Earth orbit and TLI compared to its crew variant.
>
> – SLS with a 27.6-ft. (8.4 m)-diameter payload fairing offers immense volume for large payloads, such as habitat modules, space telescopes, and interplanetary probes. Additional stages can even be encapsulated with a spacecraft to make high-energy missions to the outer planets possible.
>
> – The ultimate Block 2 design in its cargo configuration can launch more than 101,000 lbs. to TLI and more than 80,000 lbs. to Mars.

More Powerful Rockets for Future Missions

The initial SLS Block 1 rocket configuration is manifested for the first three Artemis missions to the Moon. The Block 1B variant will follow the Block 1 rocket design. The Block 1B variant incorporates several upgrades to improve SLS performance, allowing the rocket to launch larger and heavier payloads to deep space destinations.

SLS Block 1B uses the same core stage and twin five-segment solid rocket boosters as the Block 1 rocket. New-production RS-25 engines will provide slightly more thrust than the shuttle-era engines and cost 30 percent less to manufacture.

The primary increase in performance for the Block 1B rocket will come from replacing the single-engine ICPS with a new, more powerful four-engine Exploration Upper Stage (EUS). The Block 2 variant retains the core stage, RS-25 engines, and EUS but replaces the Block 1 steel case booster motor design with a lighter composite case, new propellant formulation, and other upgrades that increase overall payload performance.

SLS OVERVIEW
SLS VEHICLE-LEVEL TESTING

SLS Vehicle-Level Testing

The SLS Program conducted extensive testing on the rocket, ranging from basic structural materials testing to full-up integrated testing of the entire Artemis I SLS rocket. Element-level testing of the engines, boosters, core stage, and upper stage elements are discussed in those sections.

Avionics, the hardware and software that serve as the "brains" of the rocket, were developed and tested at Marshall. The Integrated Avionics Test Facilities team provided the structure and simulation capability to model the environments the rocket will experience during launch.

The avionics team was able to put development and flight-like hardware and software to the test in a virtual launch environment. Using specially developed software tools, the team simulated a variety of weather conditions as well as the Earth's rotation, SLS performance, and other conditions. They exercised flight avionics with engines, boosters, and Orion computer systems in combination with these additional variables. In fact, the team flew thousands of virtual missions in the lab before the Artemis I mission.

In addition to avionics testing, engineers at Marshall conducted a test campaign consisting of dozens of acoustic tests on a five-percent scale SLS model. The scale-model acoustic test program helped engineers understand how sounds and vibrations could affect SLS, its crew, and payloads in the launch environment. The small-scale model had four operational liquid engines and two solid rocket motors to represent SLS propulsion. The series also tested the water sound suppression system for the launch pad.

In another test, engineers developed algorithms that could make an F/A-18 test aircraft fly like SLS to test adaptive SLS flight control software capable of responding to abnormal conditions like higher thrust or wind gusts to see if the flight control system responds as designed.

Wind tunnel tests helped refine designs and smooth airflow in flight, determine the effects of wind before the rocket clears the launch tower, how SLS heats up as it pushes through the atmosphere, and how it responds to buffeting at or near the speed of sound. Engineers at NASA's Ames Research Center in California and NASA'S Langley Research Center in Hampton, Virgina, tested accurate scale models of the SLS crew and cargo variants in wind tunnel facilities at those centers.

In 2013, NASA conducted a series of F/A-18 research flights to test SLS prototype software, including the previously adaptive augmenting control component. Adaptive augmenting contol benefits total attitude control and increases flight control system performance when excessive tracking error is present and decreases responsiveness to undesirable frequency content. It expands the envelope under which the flight control system is capable of safely flying the vehicle.

The Systems Integration Lab at Marshall includes flight computers and avionics functionally identical to the core stage avionics to test all systems prior to flight.

SLS OVERVIEW
SLS VEHICLE-LEVEL TESTING

While the scale model acoustic test, adaptive flight control software, and wind tunnel tests were conducted relatively early in the SLS development phase, a major series of integrated vehicle tests was also conducted when the rocket was assembled in Kennedy's Vehicle Assembly Building. The Integrated Test and Check Out (ITCO) series, performed after SLS was assembled in the Vehicle Assembly Building and at the launch pad, thoroughly tested all the systems, end-to-end, prior to launch. Tests included powering up the rocket, ensuring avionics communicated with each other and with NASA's tracking system, a "wet dress rehearsal" to practice filling and draining propellant tanks with cryogenic propellant, and more. Also before launch, engineers will perform a test on the SLS rocket's flight termination system–and capability required on all rockets-and is only used if the rocket's flight has to be ended early.

For the early part of this test series, engineers used an Orion simulator instead of the Artemis I spacecraft. The Artemis I core stage was attached to the solid rocket boosters previously stacked on the mobile launcher. The launch vehicle stage adapter and ICPS were also integrated with the core stage. Because CubeSat payloads were being installed in the Orion stage adapter, an Orion stage adapter test article was stacked on the ICPS. Finally, an Orion/service module simulator was mated to the Orion stage adapter test article while Orion's launch abort system was being installed on the Artemis I Orion.

An Orion simulator mimicked the weight and weight distribution of the real spacecraft. The ITCO marked a key milestone prior to flight. The test series proved the functionality and system interoperability of flight and ground systems and provided key data on the vehicle design for future missions. During testing, the team collected data on vibrations, communications, launch pad umbilical retraction, avionics operation, and core stage and ICPS main propulsion systems, among other things.

Ames produced a visualization of SLS exhaust gas flow during liftoff from Launch Complex 39B to accurately understand how liquid and gases flowing during launch can cause various kinds of waves at the launch pad.

Marshall's Systems Integration Lab has the capabilities to allow the software team to fly thousands of virtual launches in a simulated environment that included wind, weather, and Earth's rotation.

A five-percent scale model of SLS undergoes acoustic testing in 2014 at Marshall.

19

SLS OVERVIEW
SLS VEHICLE-LEVEL TESTING

TEN ITCO TESTS TO PREPARE FOR FLIGHT:

1. **Modal Tests:** These tests characterize vehicle dynamics and include booster push/pull testing, a modal tap test, and integrated modal testing on the mobile launcher and crawler-transporter with Orion stage adapter test article and Orion mass simulator using hydraulic shakers and calibrated hammers.

2. **Interface Verification Test:** Also performed with the Orion stage adapter test article and the Orion simulator "as well as" with the Artemis I Orion, this test verifies functionality and interoperability of SLS-to-Orion interfaces.

3. **Communications End-to-End Test:** This critical test validates communications between the vehicle and tracking network.

4. **Umbilical Release and Retract Test:** Testing the timing of booster arming and firing and the command for umbilical release.

5. **Vehicle Assembly Building Project-Specific Engineering Test:** Element-level testing of SLS in the Vehicle Assembly Building.

6. **Countdown Sequence Test:** Training the launch team with Artemis I flight hardware in the loop.

7. **Flight Safety System:** Rehearsal for pre- and post-wet dress and ordnance and flight termination system communications.

8. **Dynamic Rollout Test:** Rolling the vehicle from the Vehicle Assembly Building to Launch Complex 39B and back to compare actual loads to analytical models.

9. **Pad Project-Specific Engineering Test:** At the launch pad, testing radio frequency; guidance, navigation, and control; and performing final ordnance tests.

10. **Wet Dress Rehearsal:** Testing propellant loading procedures, structural response, thermal conditioning and loading procedures; vehicle control system; avionics and software checkout; electromagnetic interference; guidance and navigation; main propulsion system; and engine and booster nozzle steering.

Modal testing underway in High Bay 3 inside the Vehicle Assembly Building at Kennedy, using the Orion stage adapter structural test article and the mass simulator for Orion. The test hardware articles have the same weight and mass characteristics as their respective flight components.

**SLS OVERVIEW
SLS VEHICLE-LEVEL TESTING**

At the conclusion of ITCO, the Orion mass simulator and Orion stage adapter test article were removed from the stack. The Artemis I Orion stage adapter, with CubeSat payloads installed, and the Orion spacecraft were placed on the integrated stack for flight.

The final major test of the ITCO series was wet dress rehearsal. For this test of the full flight stack, the crawler-transporter carried the Artemis I rocket and Orion spacecraft on the mobile launcher to Launch Complex 39B. There, the rocket – and the launch team – went through a full countdown rehearsal, including power up, systems checkout, and filling the entire rocket and spacecraft with all the propellants.

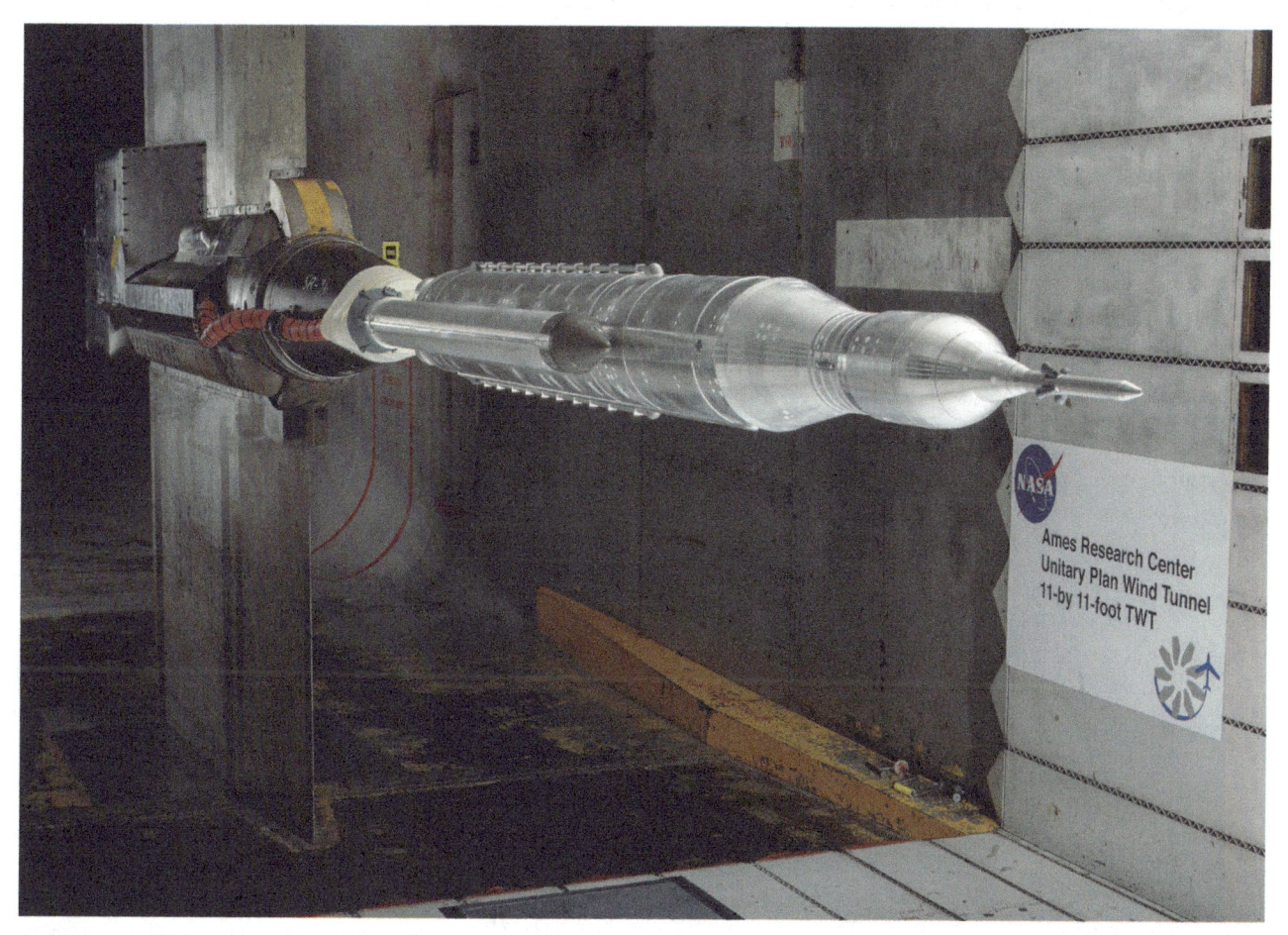

A scale model of SLS was tested in an 11-by 11-ft. (3.4-by 3.4-m) Transonic Wind Tunnel at Ames. Data from the wind tunnel tests at Ames and Langley was used to improve the design and stability of SLS.

21

SLS OVERVIEW
TRANSPORTATION, LOGISTICS,
AND PATHFINDERS

Super Heavy Lifting on the Ground: SLS Transportation, Logistics, and Pathfinders

NASA has not built a rocket on the scale of SLS, able to take astronauts and cargo to the Moon, since the Saturn V of the Apollo Program in the 1960s and 1970s.

To give transportation and ground handling crews the opportunity to rehearse and perfect logistics and handling procedures prior to receiving Artemis I hardware, the SLS Program provided several full-scale "pathfinders" to pave the way for smooth shipping, handling, and lifting operations.

The core stage pathfinder is a full-scale mockup structurally similar to the core stage in shape, size, weight, center of gravity, and handling interfaces. Lacking the internal tanks and equipment of a flight stage, the pathfinder was used to practice transport, handling, and lifting operations at Michoud, Stennis, and Kennedy before teams worked with the Artemis I core stage.

Moving flight hardware during manufacturing and testing within facilities and onto rail cars, barges, and the Super Guppy requires care and specialized equipment. These elements are too large to transport using dollies and hand trucks! While the RS-25 engines move by truck and the booster motor segments travel by rail, the core stage, ICPS, and launch vehicle stage adapter require their own transportation and handling equipment.

NASA's Super Guppy aircraft transported the Orion stage adapter to Kennedy.

The core stage pathfinder mimics the structure of the SLS core stage, providing transportation, logistics, and lifting crews with the opportunity to rehearse with full-scale hardware before handling one-of-a-kind flight hardware.

SLS OVERVIEW
TRANSPORTATION, LOGISTICS,
AND PATHFINDERS

NASA's barge Pegasus is the largest vehicle used to transport SLS elements as well as the core stage pathfinder. Pegasus ferried the Artemis I core stage from Michoud where it was manufactured, to Stennis for Green Run testing to hot fire the core stage. After Green Run, Pegasus carried the core stage to Kennedy. Pegasus also transported the launch vehicle stage adapter from Marshall to Kennedy.

Pegasus was designed and built in 1999 to transport space shuttle external tanks from Michoud to Kennedy. It replaced the Poseidon and Orion barges that were used to carry Saturn rocket stages and hardware for the Apollo Program. Pegasus was modified in 2014 to carry the longer SLS core stage. A 115-ft. (35 m) section was removed and replaced with a 165-ft. (50 m) section capable of carrying more weight and lengthening Pegasus from 260 ft. (79 m) to 310 ft. (94 m). Pegasus has no engines and instead is moved by tugboats and towing vessels.

ULA uses its R/S RocketShip, formerly the Delta Mariner, to transport the ICPS from its rocket factory in Decatur, Alabama, to the Delta Operations Center near Kennedy prior to stacking in the Vehicle Assembly Building.

For surface transportation, SLS relies on specialized Ground Support Equipment (GSE). This is a set of modular equipment ranging from smaller brackets, shackles, and pins that secure the giant rocket hardware to the slow, flatbed motorized transports that move rocket components between buildings, barge, and test stands.

Engineers and technicians at Kennedy used the 212-ft. (65 m) pathfinder core stage simulator to practice lifts and moves in the Vehicle Assembly Building.

A full-scale SLS core stage pathfinder was used by crews at Michoud, Stennis, and Kennedy to practice handling, logistics, crane lifts, and other operations before working with Artemis I flight hardware.

23

SLS OVERVIEW
TRANSPORTATION, LOGISTICS, AND PATHFINDERS

Like the flight hardware that depends on it, the GSE was designed and built to exact specifications, and its uses – and instructions for those uses – are documented in detail. Every move is carefully choreographed by the operations teams that use it.

NASA's barge Pegasus, used to transport external tanks during the Space Shuttle Program, was lengthened and strengthened to transport the SLS core stage. A 115-ft. (35 m) section of the barge was removed and replaced with a specially reinforced 165-ft. (50 m) section to increase the weight Pegasus can carry. The length of the barged increased from 260 ft. (79 m) to 310 ft. (94 m).

EGS and Jacobs technicians worked with inert booster motor segments as pathfinders in Kennedy's Rotation, Processing, and Surge Facility (RPSF) and Vehicle Assembly Building before assembling the Artemis I booster motor segments.

NASA's Super Guppy cargo aircraft transported the Orion stage adapter from Marshall, where it was manufactured, to Kennedy.

Exterior Features of SLS

Thermal Protection System

SLS has several distinctive external features. The orange color of the core stage is the thermal protection system, or spray-on foam insulation. This insulation, along with other materials such as cork, provides thermal protection for every part of the rocket, although it is most visible on the core stage. The insulation is flexible enough to move with the rocket but rigid enough to take the aerodynamic pressures as SLS accelerates from 0 to 17,400 mph (28,003 km/h) and soars to more than 100 mi. (161 km) above Earth in just 8 minutes.

The cryogenic propellant (liquid hydrogen and liquid oxygen) that powers the RS-25 engines must stay extremely cold to remain liquid. Hydrogen must remain at -423 degrees Fahrenheit (-253 degrees Celsius) and oxygen at -297 degrees Fahrenheit (-183 degrees Celsius). If temperatures rise too high, the propellant will become a gas.

Materials engineers qualified the third-generation, orange-colored spray-on foam insulation to meet the harsh environments that the SLS will experience. At the same time, they made the foam more environmentally friendly. When the foam is applied, it gives the rocket a light-yellow color that the Sun's ultraviolet rays eventually "tan," giving the SLS core stage its signature orange color.

The SLS gets its signature orange color when its spray-on foam insulation darkens with exposure to sunlight. The core stage below the work platform was exposed to months of sunlight during Green Run testing at Stennis, while the launch vehicle stage adapter being lowered into place above the platform has been stored indoors since its completion.

SLS OVERVIEW
EXTERIOR FEATURES OF SLS

Livery and Photogrammetric Markings

In addition to the SLS core stage's distinctive orange color, several other markings are visible on the outside of the rocket. The most visible are the national and agency livery markings such as "USA," the U. S. flag, the NASA "meatball" and "worm" marks, and the ESA mark.

SLS, as well as the Orion crew spacecraft and the mobile launcher, also have black-and-white markings that play an important role in Artemis I. These photogrammetric markings are the black and white checkerboards, squares, and circles on the outside of the rocket. These markings serve as imagery references for engineering photo and video documentation of SLS attitude and position relative to ground structure during liftoff and ascent. Markings range in size from 0.2 in. (0.5 cm) to 3 x 3 ft. (0.9 x 0.9 m). On the solid rocket boosters, some of the markings are a multi-checkered pattern measuring 24 x 130 in. (0.6 m x 3.3 m).

Some of the markings are visible only to internal cameras and will capture separation and jettison events. The launch vehicle stage adapter has eight internal markings; the ICPS has 12; the forward bay of the European Service Module has 30; the spacecraft adapter has 12; and spacecraft adapter jettison panel 1 has 12 markings.

Engineers are interested in every movement of the rocket, both independently and in relation to the mobile launcher, in every phase of flight, from the launch pad to its return to Earth. Most of the markings will be used during separation events, such as SLS from the mobile launcher, boosters from the core stage, core stage/launch vehicle stage adapter from the ICPS, and ICPS/Orion stage adapter from Orion and its service module.

The imagery markings help to characterize these movements, and the data will feed back into computer models to understand actual vs. predicted movements. Every key launch event and its marking were analyzed for placement. Key considerations were where to place onboard cameras to survive the harsh launch environment, as well as where ground cameras are located.

SLS has four cameras on the engine section that look forward, two on the intertank looking aft, and two inside the launch vehicle stage adapter to capture the ICPS separation event. The engine section cameras are forward-facing and have a tilt of 10 degrees and a pan of 0 degrees. The aft-facing intertank cameras have a tilt of 2.5 degrees and a pan of 5 degrees. There are more than 150 ground cameras used for inspecting or monitoring the vehicle during launch. A few of those are for public affairs purposes, but most of them provide imagery to engineers. For Artemis I, NASA will also have one WB-57 aircraft capturing aerial imagery. There are also plans for other airborne and ship-based imagery during re-entry and recovery of Orion.

Technicians with lead solid rocket boosters contractor Northrop Grumman paint photogrammetric markings on one segment of an SLS solid rocket motor. The black-and-white targets painted on each element of the rocket provide targets for cameras, which then use those positions compared with fixed positions on the ground and on the mobile launcher to precisely locate SLS in space.

SLS OVERVIEW
EXTERIOR FEATURES OF SLS

Instrumentation and Rocket Telemetry

As a test flight, the basic Artemis I mission is to successfully confirm performance and operational capability of the integrated flight and ground elements in performing the lunar mission, including for SLS countdown, launch, ascent, and putting Orion on course to the Moon. Success will be measured by flight data collected on the ground and in flight. Much of that flight data will be generated by computers on the engines, boosters, core stage, ICPS, and thousands of sensors attached to the rocket.

That data and instrumentation to capture the data is divided into three categories: operational flight instrumentation, engineering flight instrumentation, and development flight instrumentation. Operational data is hardware and/or software required to operate the vehicle. Engineering data is hardware or software data mainly used for post-flight performance analysis. Developmental data is used to increase confidence in the computer models – predictions – of how the vehicle should perform.

SLS will transmit data from more than 27,000 sources of data to the ground for analysis. Of those, more than 20,000 channels will be operational and engineering data with the rest devoted to development data. Each element – core stage, engines, boosters, payload, command telemetry, and imagery – has a specified allocated bandwidth for transmission during flight.

All this data will be collected and analyzed to ensure SLS is the most reliable, highest-performing, and safest rocket that NASA has ever sent to deep space. It is part of the Artemis I test flight philosophy. Engineers will rigorously test all the systems in deep space, because Artemis I is just the beginning. Following Artemis I, crewed flights will commence, culminating with the first woman and the first person of color walking on the Moon.

Part of the Kennedy Uplink Station, the S-band ground tracking system at Kennedy provides crucial tracking and communication capabilities following liftoff of SLS and Orion.

THE ROLE OF SLS IN LAUNCHING ARTEMIS I

THE ROLE OF SLS IN LAUNCHING ARTEMIS I

Artemis I will provide the first integrated flight test of SLS, Orion, and ground processing and launch facilities at Kennedy. For Artemis I, SLS will launch an uncrewed Orion to a distant retrograde orbit about the Moon in a flight test designed to verify vehicle models, manufacturing, and operations. The data engineers receive from Artemis I on the vehicle's performance will be valuable for future flights with astronauts.

The Artemis I countdown will begin with a "call to stations" for engineers and technicians working operations consoles at the Launch Control Center (LCC) at Kennedy, the Mission Control Center at Johnson, the SLS Engineering Support Center (SESC) at Marshall, and several contractor locations.

Artemis I: SLS Flight Profile

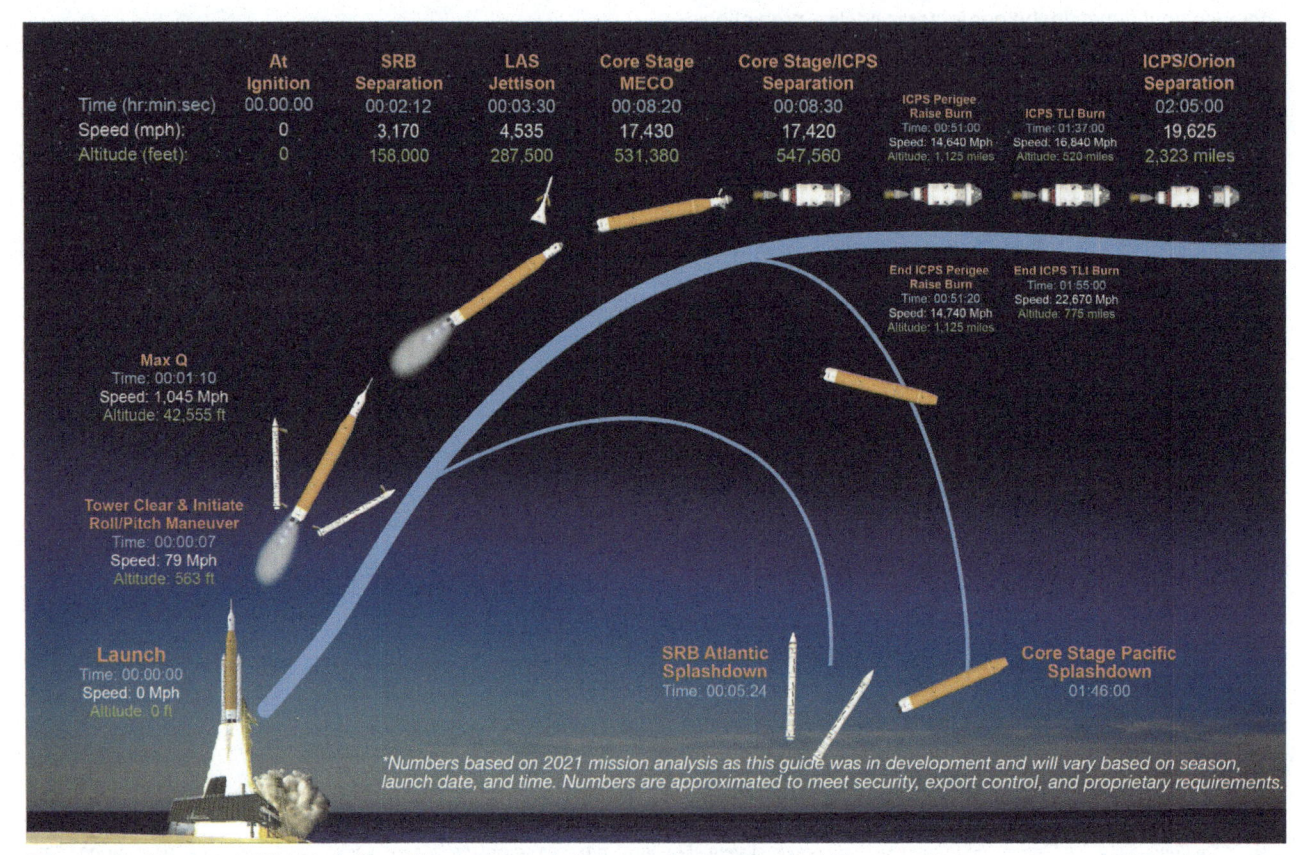

	At Ignition	SRB Separation	LAS Jettison	Core Stage MECO	Core Stage/ICPS Separation			ICPS/Orion Separation
Time (hr:min:sec)	00:00:00	00:02:12	00:03:30	00:08:20	00:08:30	ICPS Perigee Raise Burn Time: 00:51:00 Speed: 14,640 Mph Altitude: 1,125 miles	ICPS TLI Burn Time: 01:37:00 Speed: 16,840 Mph Altitude: 520 miles	02:05:00
Speed (mph):	0	3,170	4,535	17,430	17,420			19,625
Altitude (feet):	0	158,000	287,500	531,380	547,560			2,323 miles

End ICPS Perigee Raise Burn
Time: 00:51:20
Speed: 14,740 Mph
Altitude: 1,125 miles

End ICPS TLI Burn
Time: 01:55:00
Speed: 22,670 Mph
Altitude: 775 miles

Max Q
Time: 00:01:10
Speed: 1,045 Mph
Altitude: 42,555 ft

Tower Clear & Initiate Roll/Pitch Maneuver
Time: 00:00:07
Speed: 79 Mph
Altitude: 563 ft

Launch
Time: 00:00:00
Speed: 0 Mph
Altitude: 0 ft

SRB Atlantic Splashdown
Time: 00:05:24

Core Stage Pacific Splashdown
01:46:00

*Numbers based on 2021 mission analysis as this guide was in development and will vary based on season, launch date, and time. Numbers are approximated to meet security, export control, and proprietary requirements.

Major events of the Artemis I flight, including solid rocket booster jettison, core stage and launch vehicle stage adapter separation, and ICPS and Orion stage adapter separation from Orion. Following separation with Orion, CubeSat payloads will be released into deep space.

THE ROLE OF SLS IN LAUNCHING ARTEMIS I

For the flight, SLS will lift off, and the solid rocket boosters will burn through their propellant and separate after approximately two minutes. The core stage and RS-25s will deplete propellant after approximately six more minutes. After main engine cutoff, the ICPS, Orion stage adapter, and Orion will separate from the core stage and launch vehicle stage adapter. Engineers on the ground will verify the health of ICPS and Orion systems before giving the approval for the TLI burn by the ICPS. The TLI maneuver precisely targets a point about the Moon that will guide Orion close enough to be captured by the Moon's gravity.

Following the TLI burn, the ICPS separates from Orion. Orion's service module propels it from this point, making course correction burns as needed to begin its three-week mission. Its mission almost complete, the upper stage, meanwhile, continues on a similar path, deploying several CubeSats along the way to study the Moon or head farther out to deep space destinations.

The boosters will fall into the Atlantic Ocean about 140 mi. (225 km) off the Florida coast, north of the Bahamas. The core stage/launch vehicle stage adapter will fall into the Pacific Ocean east of Hawaii, west of Baja, California. After Orion separation, the ICPS will enter heliocentric (around the Sun) orbit.

Flight testing mitigates risk by collecting flight data to further anchor engineering models used to certify the design. Each program – EGS, Orion, and SLS – has objectives for the Artemis I flight test mission.

Artemis I is a test flight to verify SLS vehicle models, manufacturing, and operations. The measure of mission success will be the data returned on the vehicle and its performance.

ARTEMIS I OBJECTIVES SPECIFIC TO SLS:
- Testing pre-launch operations.
- Collecting flight telemetry and imagery to evaluate structural integrity of the vehicle from ascent through ICPS/Orion separation.
- Testing the core stage flight computers' ability to run an autonomous launch sequence.
- Initiation of hydrogen burn-off igniters.
- Guidance and navigation.
- Core stage engine ignition, throttling, steering, and cutoff.
- Booster ignition, separation, and ocean impact.
- Mobile launcher umbilical separation.
- Liftoff and launch tower clearance.
- Performance of a heads-down roll maneuver.
- Transonic, maximum dynamic pressure, and maximum acceleration.
- Testing communication systems.
- Core stage/ICPS staging and core stage ocean impact.
- ICPS operations, including steering, engine start and shutdown, perigee raise and TLI engine burns, and disposal maneuver.
- Orion/ICPS staging.
- Secondary payload deployment.

Technicians installed the liquid hydrogen tank structural test article in its test stand at Marshall at night when winds were minimal.

ARTEMIS I SECONDARY PAYLOADS

Artemis I Secondary Payloads

The unmatched SLS payload mass capability and the unused volume of the Orion stage adapter provide a rare opportunity for small, low-cost science and technology experiments to be deployed into deep space.

These secondary payloads are not much larger than a shoebox but contain science and technology investigations or technology demonstrations that help pave the way for future deep space human exploration. International space agency partners and universities are involved with several of the CubeSat payloads.

The Artemis I CubeSats are 6U in size. One U – or unit – is 10 cm x 10 cm x 10 cm. The Artemis I payloads are limited to about 30 lbs. (14 kg) each. Several of the CubeSats chosen to fly on Artemis I are lunar-focused and may help NASA address Strategic Knowledge Gaps to inform research strategies and prioritize technology development of human and robotic exploration.

Other missions will be testing innovative propulsion technologies, studying space weather, analyzing the effects of radiation on organisms, and providing high-resolution imagery of Earth and the Moon.

The NEA Scout CubeSat payload is integrated into its dispenser prior to the Artemis I flight.

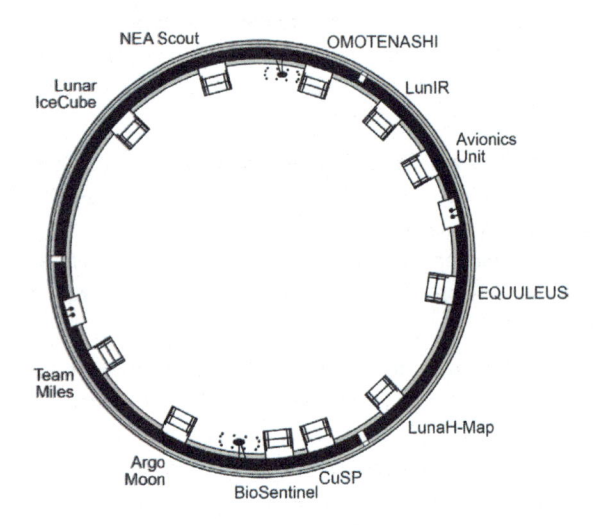

Payload berth locations on the Orion stage adapter.

ARTEMIS I SECONDARY PAYLOADS

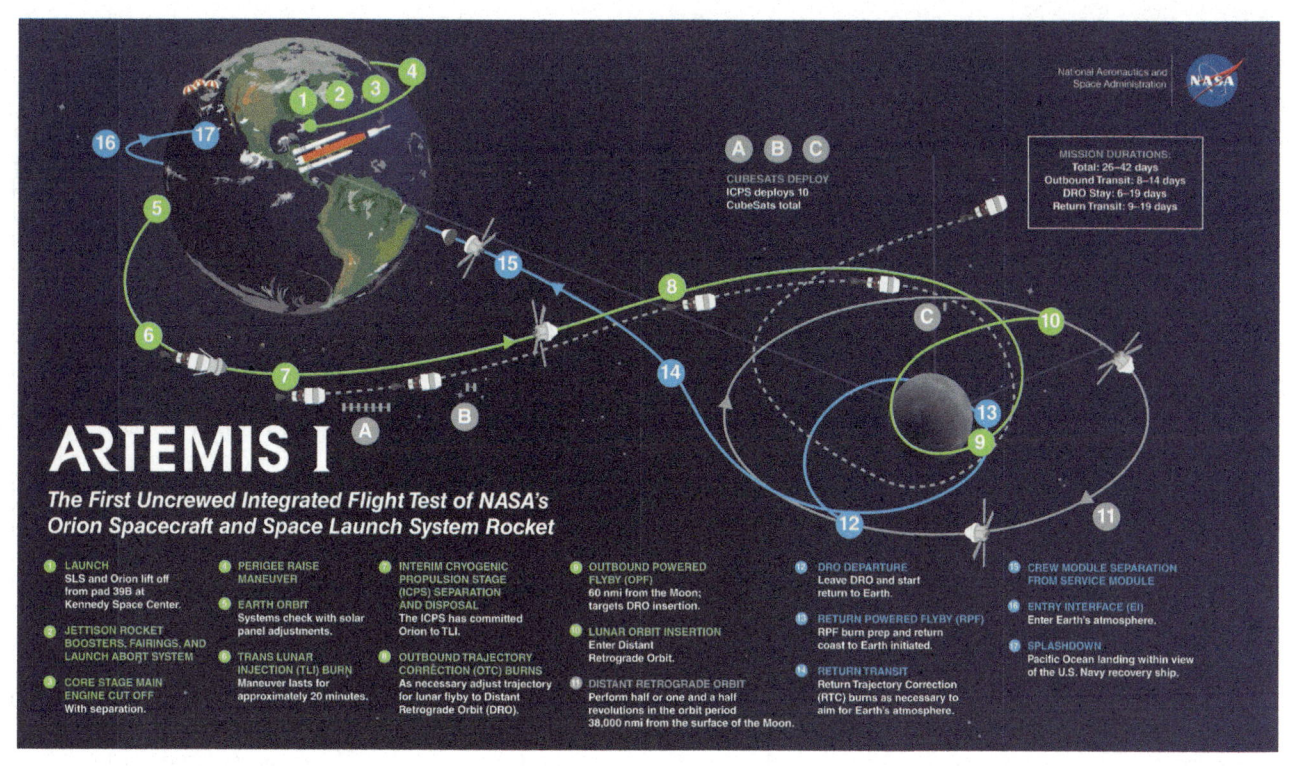

ARTEMIS I
The First Uncrewed Integrated Flight Test of NASA's Orion Spacecraft and Space Launch System Rocket

MISSION DURATIONS:
Total: 26–42 days
Outbound Transit: 8–14 days
DRO Stay: 6–19 days
Return Transit: 9–19 days

CUBESATS DEPLOY
ICPS deploys 10 CubeSats total

1. **LAUNCH** — SLS and Orion lift off from pad 39B at Kennedy Space Center.
2. **JETTISON ROCKET BOOSTERS, FAIRINGS, AND LAUNCH ABORT SYSTEM**
3. **CORE STAGE MAIN ENGINE CUT OFF** — With separation.
4. **PERIGEE RAISE MANEUVER**
5. **EARTH ORBIT** — Systems check with solar panel adjustments.
6. **TRANS LUNAR INJECTION (TLI) BURN** — Maneuver lasts for approximately 20 minutes.
7. **INTERIM CRYOGENIC PROPULSION STAGE (ICPS) SEPARATION AND DISPOSAL** — The ICPS has committed Orion to TLI.
8. **OUTBOUND TRAJECTORY CORRECTION (OTC) BURNS** — As necessary adjust trajectory for lunar flyby to Distant Retrograde Orbit (DRO).
9. **OUTBOUND POWERED FLYBY (OPF)** — 60 nmi from the Moon; targets DRO insertion.
10. **LUNAR ORBIT INSERTION** — Enter Distant Retrograde Orbit.
11. **DISTANT RETROGRADE ORBIT** — Perform half or one and a half revolutions in the orbit period 38,000 nmi from the surface of the Moon.
12. **DRO DEPARTURE** — Leave DRO and start return to Earth.
13. **RETURN POWERED FLYBY (RPF)** — RPF burn prep and return coast to Earth initiated.
14. **RETURN TRANSIT** — Return Trajectory Correction (RTC) burns as necessary to aim for Earth's atmosphere.
15. **CREW MODULE SEPARATION FROM SERVICE MODULE**
16. **ENTRY INTERFACE (EI)** — Enter Earth's atmosphere.
17. **SPLASHDOWN** — Pacific Ocean landing within view of the U.S. Navy recovery ship.

The Artemis I test flight includes objectives such as testing SLS performance, Orion's heat shield at lunar re-entry velocities, testing systems in deep space, exercising countdown and mission operations, and deploying CubeSat payloads in deep space.

The Jacobs team at NASA Kennedy installing the last of 10 CubeSats in the SLS rocket's Orion stage adapter. BioSentinel, the final CubeSat to be loaded, will study how radiation affects living organisms in deep space.

ARTEMIS I SECONDARY PAYLOADS

The CubeSats will be deployed after Orion separates and is a safe distance away. Each payload will be ejected with a spring mechanism from dispensers installed on the Orion stage adapter. Artemis I manifested payloads are listed below with their provider, area of exploration, and a brief summary:

Payload	Developer(s)	Sponsor	Destination	Mission
ArgoMoon	Argotec	Agenzia Spaziale Italiana	Geocentric orbit with high eccentricity and apogee close to the Moon	Photograph the ICPS, CubeSat deployment, the Earth and Moon using HD cameras and advanced imaging software
BioSentinel	NASA Ames, NASA Johnson, Loma Linda University Medical Center, University of Saskatchewan	NASA Advanced Exploration Systems (AES)	Heliocentric orbit via lunar flyby	Use yeast as a biosensor to evaluate the effects of ambient space radiation on DNA
CubeSat to Study Solar Particles (CuSP)	Southwest Research Institute, NASA Goddard	NASA Science Mission Directorate (SMD)	Deep space	Study the sources and acceleration mechanisms of solar and interplanetary particles in near-Earth orbit
EQUilibriUm Lunar-Earth point 6U Spacecraft (EQUULEUS)	University of Tokyo	Japanese Aerospace Exploration Agency (JAXA)	Earth-Moon L2 point	Demonstrate trajectory control techniques within the Sun-Earth-Moon region and image Earth's plasmasphere
Lunar IceCube	Morehead State University, NASA JPL, NASA Goddard, BUSEK	NASA Next Space Technologies for Exploration Partnerships (NextSTEP)	Lunar orbit	Search for water (and other volatiles) in ice, liquid, and vapor states using infrared spectrometer
Lunar-Polar Hydrogen Mapper (LunaH-Map)	Arizona State University	NASA SMD	Lunar orbit	Perform neutron spectroscopy to characterize abundance of hydrogen in permanently shaded craters
LunIR	Lockheed Martin Space Systems	NASA NextSTEP	Heliocentric orbit via lunar flyby	Use a miniature high-temperature Mid-Wave Infrared (MWIR) sensor to characterize the lunar surface
Near Earth Asteroid (NEA) Scout	NASA Marshall	NASA AES	NEA within ~1.0 astronomical unit (AU) of Earth	Detect target NEA, perform reconnaissance and close proximity imaging
Outstanding MOon exploration Technologies demonstrated by Nano Semi-Hard Impactor (OMOTENASHI)	Institute of Space and Astronautical Science (ISAS)/JAXA	JAXA	Lunar surface	Develop world's smallest lunar lander and observe lunar radiation environment
Team Miles	Miles Space, LLC	NASA Cube Quest Challenge, sponsored by NASA Space Technology Mission Directorate's Centennial Challenges	Deep space	Demonstrate propulsion using plasma thrusters; compete in NASA's Deep Space Derby

ARTEMIS I SECONDARY PAYLOADS

For Artemis I, several "bus stops" were defined to provide payload developers with reference points when deciding when to deploy from the vehicle. For example, bus stop 1 is the first opportunity to deploy after Orion has separated and the ICPS has completed its mission operations. Bus stop 1 also coincides with the entrance to the second Van Allen Belt. Bus stop 2 is about one hour past the Van Allen Belts to reduce radiation exposure to the CubeSats. Bus stop 3 is equidistant between the Earth and Moon. Some payloads are deploying between bus stops.

Artemis I CubeSat	Mission	Estimated Deployment Time	Mission Orbit	Mission Duration	Disposal
ArgoMoon	Perform proximity operations with the ICPS post-disposal; take external imagery of Earth and Moon by testing an advanced software imaging recognition system with high-definition cameras	L + ~ 3 hrs. + 40 min.	Geocentric	180 days	Heliocentric orbit
CuSP	Study solar and interplanetary particles supporting space weather research by determining proton radiation levels during Solar Energetic Particle (SEP) events and identifying suprathermal properties that could help predict geomagnetic storms	L + ~ 8 hrs. + 3 min.	Heliocentric	75 days	Heliocentric orbit
LunaH-Map	Research to understand the quantity of hydrogen-bearing materials in cold traps in permanently shaded craters of the Moon's South Pole	L + ~ 5 hrs. + 33 min.	Lunar	601 days (1.6 years)	Lunar surface
LunIR	Perform a lunar flyby using a miniature high-temperature Mid-Wave Infrared (MWIR) sensor collecting spectroscopy and thermography data to address questions related to surface characterization, remote sensing, and site selection for future missions	L + ~ 6 hrs. + 3 min.	Lunar flyby	30 days	Heliocentric orbit
BioSentinel	Contains a yeast radiation biosensor that will measure effects of space radiation on DNA	L + ~ 3 hrs. + 40 min.	Lunar flyby	1 year	Heliocentric orbit
Lunar IceCube	Search for water in ice, liquid, and vapor forms as well as other lunar volatiles using a compact infrared spectrometer	L + ~ 3 hrs. + 40 min.	Lunar	< 2 years	Lunar surface
NEA Scout	Equipped with a solar sail to rendezvous with an asteroid, gather detailed imagery, and observe the asteroid's position in space	L + ~ 5 hrs. + 10 min.	Earth and lunar flybys, targeting cislunar escape, enabling rendezvous with near-Earth asteroid	2.5 years	Heliocentric orbit
OMOTENASHI	Land on the lunar surface to demonstrate the feasibility of the hardware for distributed cooperative exploration systems	L + ~ 3 hrs. + 40 min.	Lunar	5 days	Lunar surface
EQUULEUS	Fly to a libration orbit around the Earth-Moon L2 point and demonstrate trajectory control techniques within the Sun-Earth-Moon region	L + ~ 3 hrs. + 40 min.	Lunar	270 days	Heliocentric orbit
Team Miles	Fly autonomously using a sophisticated onboard computer system with propulsion supplied by evolutionary plasma thrusters	L + ~ 7 hrs. + 3 min.	Heliocentric	1 year max.	Heliocentric orbit

THE ELEMENTS OF SLS

THE ELEMENTS OF SLS

Each major element of SLS – the core stage, RS-25 engines, solid rocket boosters, ICPS, Launch vehicle stage adapter, and Orion stage adapter – serves a unique purpose. Each element is built and tested by a prime contractor working with numerous subcontractor suppliers. As part of the national strategic infrastructure, SLS is built by skilled American engineers and technicians at small and large contractor locations across the United States.

THE ELEMENTS OF SLS
CORE STAGE

Core Stage

The SLS core stage is the tallest rocket stage NASA has ever built. It measures approximately 212 ft. (64.6 m) tall and 27.6 ft. (8.4 m) in diameter (excluding thermal protection system foam and flanges). Its fully fueled weight, excluding engines, is 2.4 million lbs. (1,088 t).

The SLS core stage contains four RS-25 engines, their liquid hydrogen and liquid oxygen propellant supply, and the avionics and software that control SLS operation and flight until the core stage separates from the ICPS. It is literally the core of SLS, supporting other stages, spacecraft, and payloads atop its uppermost section and serving as the attach point for the two solid rocket boosters.

The core stage is designed to operate for the entire roughly 480-second launch from ground to Earth orbit, reaching speeds of nearly Mach 23 and more than 530,000 ft. (161.5 km) in altitude before it separates from the ICPS, Orion stage adapter, and Orion spacecraft.

The core stage is made up of 10 major barrel sections, four dome sections, and seven rings. Each cylindrical barrel section consists of eight aluminum panels which vary in length and height depending on the section. Those panels are friction-stir welded or bolted vertically and horizontally to form the five major sections of the core stage:

– Engine section.
– Liquid hydrogen tank.
– Intertank.
– Liquid oxygen tank.
– Forward skirt.

The intertank is the lone bolted section to provide added strength to carry booster loads.

The core stage is a major new development for the SLS Program, while other key elements, such as the RS-25 engines, solid rocket booster structures, and ICPS have previous spaceflight experience.

A key goal of the SLS design was to reuse space shuttle components or design within shuttle heritage experience where possible. For example, the core stage has the same diameter as the shuttle external tank and the propellant feedlines, and fill and drain ducts were sized around heritage joints and existing valves.

Beginning at the bottom, or aft end, of the stage, the engine section houses four RS-25 main engines, the engine thrust structure, propellant ducts, various avionics systems, engine thrust vector control systems, and serves as the aft attach point for the two solid rocket boosters.

The SLS core stage, the tallest rocket stage NASA has ever built, is lowered between the Artemis I boosters in the Vehicle Assembly Building.

THE ELEMENTS OF SLS
CORE STAGE

The engine section consists of a single barrel section and is 27.6 ft. (8.4 m) in diameter, 22.5 ft. (6.9 m) tall, and consists of welded aluminum isogrid panels. An aerodynamic boat tail fairing at the bottom channels airflow and protects the engines from extreme temperatures during launch.

The liquid hydrogen fuel tank is 27.6 ft. (8.4 m) in diameter and 130 ft. (39.7 m) tall. It consists of five welded barrel sections each 22 ft. (6.7 m) tall and two end domes. The aft end of the liquid hydrogen tank includes four liquid hydrogen feedlines to the RS-25 engines. The tank has a capacity of 537,000 gal. (2 million L) of liquid hydrogen.

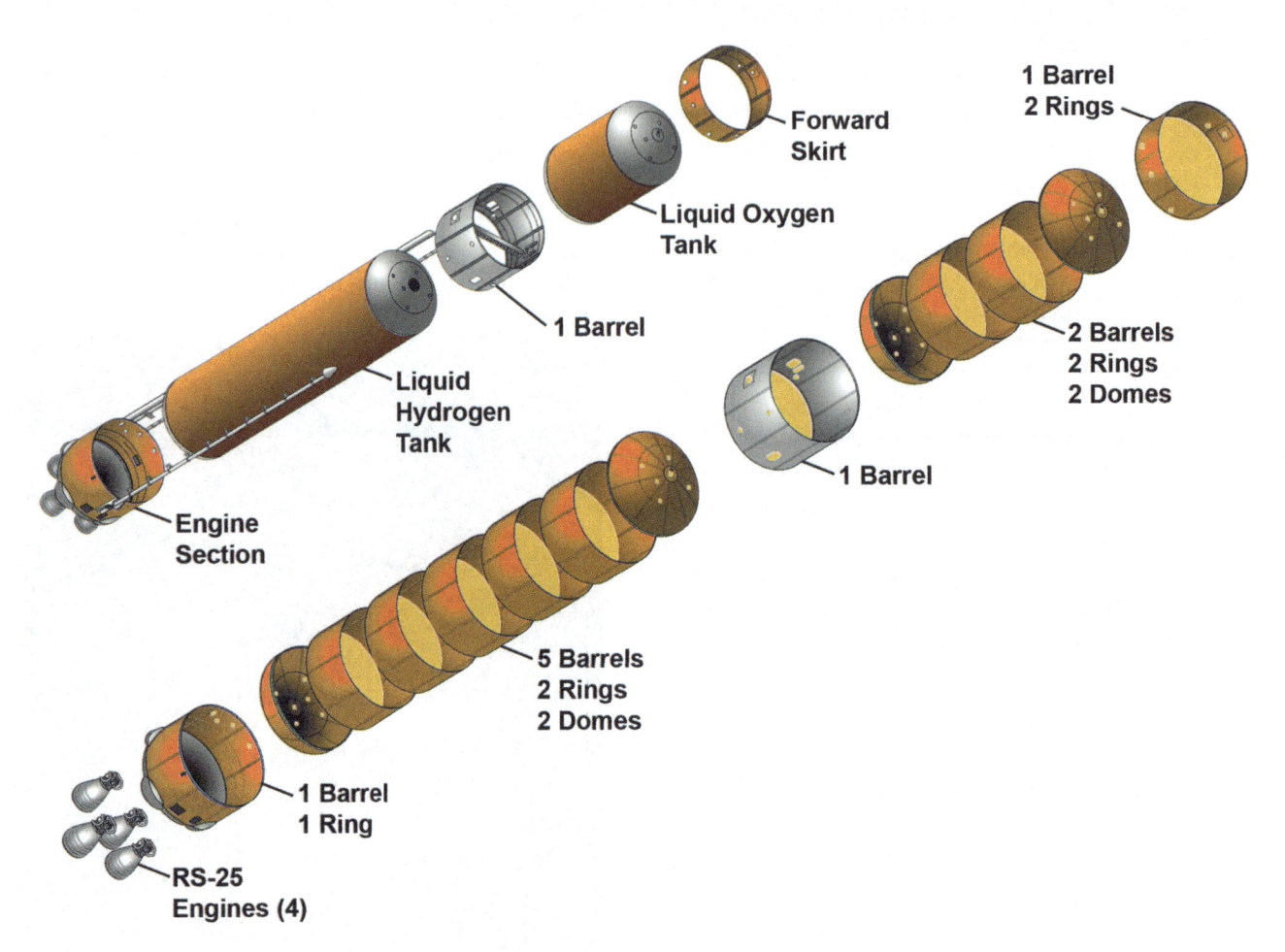

The SLS core stage is comprised of 10 barrel sections, four dome sections, and seven rings as shown in this illustration of the core stage major components.

THE ELEMENTS OF SLS CORE STAGE

The SLS Artemis I engine section, left, and liquid hydrogen tank, right, at Michoud prior to integration.

The intertank separates the upper hemispherical dome of the liquid hydrogen tank from the lower hemispherical dome of the liquid oxygen tank and serves as the forward attach point for the boosters. The intertank measures 27.6 ft. (8.4 m) in diameter and 21.8 ft. (6.7 m) tall. It contains a thrust structure to carry loads imparted from the solid rocket boosters during ascent. The intertank also contains several avionics components, including two rear-facing cameras and interfaces to the liquid hydrogen and liquid oxygen tanks.

The liquid oxygen tank is 27.6 ft. (8.4 m) in diameter and 51 ft. (15.6 m) tall. It consits of two 15.6-ft. (4.8 m) barrel sections of isogrid aluminum panels and two domes. Liquid oxygen is fed to the engine section and engines through a pair of "downcomer" ducts that exit the intertank on opposite sides and run down the core stage. The liquid oxygen tank has a capacity of 196,000 gal. (741,941 L). The thermal protection system on the tank minimizes boiloff of the -297 degrees Fahrenheit (-182.8 Celsius) liquid oxygen. Gaseous oxygen is vented overboard.

The SLS Artemis I intertank at Michoud, prior to integration between the propellant tanks, has forward attach points for the two solid rocket boosters.

THE ELEMENTS OF SLS
CORE STAGE

The forward skirt is located at the top of the core stage. It connects the core stage to the Integrated Spacecraft/Payload Element. The aluminum isogrid structure is 27.6 ft. (8.4 m) in diameter and 10.4 ft. (3.2 m) tall. It houses the majority of the vehicle's avionics and has connections to launch pad utility umbilicals, the vehicle stabilization system that helps secure SLS to the mobile launcher, access doors, vent system, pressurant lines, and antennas.

The main propulsion system in the core stage consists of the ducts, valves, and other equipment that supply and control the flow of liquid hydrogen and liquid oxygen propellants, as well as gaseous helium and nitrogen pressurants for valve actuation and line/volume purges. To accomplish those functions, the main propulsion system has four subsystems:

- Liquid oxygen.
- Liquid hydrogen.
- Pressurization.
- Pneumatics supplied by ground systems prior to launch.

Major drivers in the design of the main propulsion system included the main propellant tank configuration, main engine configuration, reliability and affordability, mission requirements, and component mounting. For instance, the main propulsion system flow rates and interfaces were designed around the RS-25 configuration and the need to supply propellants to the engines under temperature and pressure conditions required by the engines. The orientation of the engine hydrogen and oxygen feedlines in the engine section determined the feed system layout for the main propulsion system.

The core stage forward skirt contains the flight computers for the Block 1 SLS rocket.

Connecting feedlines and other external components to the fully assembled Artemis I core stage at Michoud.

The liquid oxygen tank being manufactured for the Artemis I flight at Michoud.

THE ELEMENTS OF SLS
CORE STAGE MANUFACTURING, TEST, AND CHECKOUT

Manufacturing, Test, and Checkout

The SLS core stage is manufactured at Michoud, where the Saturn rocket stages used during the Apollo Program and the space shuttle's external tanks were manufactured. Boeing, as the prime contractor for the SLS core stage, is responsible for the design, manufacturing, and testing of the stage.

Hydraulic cylinders applied millions of pounds of force to the liquid oxygen tank test article, completing 20 tests until it failed, providing engineers with valuable data on the tank's structural capabilities.

The test version of the SLS liquid hydrogen tank withstood more than 260 percent of expected flight loads over five hours before it buckled, then ruptured, giving engineers real-world data to reinforce computer models.

A significant development campaign preceded and ran concurrently with manufacturing of flight core stage components. Hundreds of metal "coupons" followed by test panels were welded at Marshall to develop the friction-stir welding process used to manufacture the core stage components at Michoud.

Full-size core stage barrel sections and tanks were welded using production tools at Michoud. Those included production of full-size flight-like structural test articles for the engine section, liquid hydrogen tank, intertank, and liquid oxygen tank.

These were heavily instrumented with a range of sensors and installed in new test stands and fixtures at Marshall. They were subjected to flight loads created by hydraulic actuators applying millions of pounds of force to verify predicted performance and establish a margin of safety. The core stage structural tests, combined with testing of the SLS upper stage and adapters, was the largest structural test campaign at Marshall since tests conducted for the Space Shuttle Program more than 30 years ago. During the test campaign, five structural test articles underwent 199 separate test cases, and more than 421 gigabytes of data were collected to verify data from computer models used to design the rocket.

THE ELEMENTS OF SLS
CORE STAGE MANUFACTURING, TEST, AND CHECKOUT

Likewise, avionics and flight software, the "brains" of SLS, were developed, tested, and virtually flown thousands of times at Marshall's Integrated Avionics and Software and Systems Integration laboratories before the flight hardware and software were installed in the Artemis I core stage.

In addition, Artemis I core stage suppliers tested individual components before shipping to Michoud. Once at Michoud, Boeing tested the components and functions before installing them on the core stage. The completed stage underwent basic electrical and pneumatic testing before shipment to Stennis for the Green Run test campaign. "Green run" describes the first operational test of a new component.

At Stennis, teams from Boeing, SLS, and Stennis used the Green Run test series of progressively more flight-like tests to activate stage systems, culminating in a hot fire of the engines that simulated launch and flight of the Artemis I core stage.

Beginning in early 2020, the Artemis I core stage was installed in the B-2 test stand, used previously to support Saturn and shuttle propulsion tests. During the year it was subjected to a series of nine major Green Run tests:

– Modal (vibration).
– Core stage avionics power on and checkout.
– Safing checks (how to shut down all core stage systems in case of an anomaly).
– Main propulsion system (stage propellant valves and ducts) and RS-25 leak checks and functional tests.
– Hydraulic and thrust vector control (engine steering) checkout.
– Simulated countdown.
– Countdown and wet dress rehearsal (full stage operation, including flowing propellants to the engines for chilldown).
– Countdown and hot fire (limited duration test up to thrust vector control gimbal).
– Countdown and hot fire (full test of the stage including engines in flight-like operation).

At the end of testing, the stage was inspected and refurbished before shipping to Kennedy onboard NASA's barge Pegasus.

Eight major "Green Run" tests at Stennis of the Artemis I core stage and RS-25 engines culminated in a full-duration hot firing, the largest test of liquid-fueled engines since the Apollo Program.

THE ELEMENTS OF SLS
CORE STAGE FUN FACTS

Core Stage Fun Facts

- There are 562 cables in the core stage. The largest number – 231 – are located in the engine section.
- There are 45 mi. of cabling in the core stage and more than 18 mi. in the engine section alone.
- There are 775 independent sensors that have wire routing to them.
- There are approximately 100,000 clamps and ties securing wires and cables throughout the core stage.
- The liquid hydrogen tank shrinks about 6 in. (15.24 cm) in length and 1 in. (2.54 cm) in diameter when filled with cryogenic fuel.
- The liquid oxygen tank shrinks approximately 1.5 in. (3.81 cm) in length and 0.5 in. (1.27 cm) in diameter when filled with cryogenic propellant.
- To compensate for those changes, everything that attaches to them – ducts, vent lines, systems tunnel, brackets, etc. – must be connected by accordion-like bellows sections, slotted joints, telescoping sections, or ball joint hinges.
- Roughly 14,500 fasteners need to be drilled and filled in the intertank.
- The flight computer uses the same class microprocessor as the original Macintosh PowerBook G3. This is somewhere between a Pentium II and Pentium III. However, the operating system is much more sophisticated.
- Each flight computer is rated to operate over a temperature range of -11 to 97.7 degrees Fahrenheit (-24 to 36.5 degrees Celsius).

**THE ELEMENTS OF SLS
RS-25 ENGINES**

RS-25 Engines

Four RS-25 engines power SLS for its eight-minute climb to Earth orbit, together with a pair of solid rocket boosters that operate for the first two minutes of ascent. The RS-25 is formerly known as the Space Shuttle Main Engine (SSME). It flew successfully on 135 space shuttle missions.

The RS-25s have accumulated more than 3,000 starts and one million seconds of ground and flight hot-fire experience. The RS-25, manufactured by Aerojet Rocketdyne, is the most efficient rocket engine in its class, allowing heavier payloads to be carried without increasing the rocket's size.

The RS-25 is a staged-combustion cycle engine that produces approximately 500,000 lbs. of thrust. Each engine has an onboard computer that automatically controls start and shutdown, thrust ranging from 65 percent to 109 percent, propellant mixture ratio, and engine health.

During ascent, the engines are gimballed through two planes by hydraulic actuators to produce vehicle pitch, yaw, and roll. Several key design features that contribute to the RS-25's high performance:

- Its cyrogenic liquid hydrogen and liquid oxygen propellants, which are more efficient than hydrocarbon engines.

- Its staged-combustion operating cycle, in which propellants are burned twice, first in preburners then in the main combustion chamber, and its two high-pressure turbopumps (liquid oxygen and liquid hydrogen) fed by two low-pressure pumps, creating higher main combustion chamber pressures.

It was that power, efficiency, and reliability – as well as the knowledge and experience base – that led to selection of the RS-25 for SLS. Each RS-25 is roughly 14 ft. (4.3 m) tall, 8 ft. (2.4 m) in diameter, and weighs approximately 7,750 lbs. (3.5 t). Unlike the space shuttle, SLS will not reuse its RS-25s as the core stage size, as well as altitude and speed at MECO, make recovery impractical without significant sacrifice in payload-carrying capability. SLS is designed to launch the most ambitious space exploration missions and requires maximum performance.

The RS-25s have been upgraded with new controllers that were hot-fire tested prior to the Artemis I flight.

THE ELEMENTS OF SLS
RS-25 ENGINES

The first four SLS missions will be powered by 14 flight-proven engines and two new engines assembled from shuttle-era engine components. Upgrades to the engines include development of new engine controllers and software and the addition of nozzle insulation to protect them from booster exhaust due to the engine and booster nozzles being located roughly in the same plane. New RS-25 engines now in manufacturing will be available for the fifth and subsequent missions.

The operations of the SLS RS-25s differs from operations during the Space Shuttle Program. Each shuttle mission used three engines, while each SLS mission uses four. Shuttles routinely operated with SSMEs throttled to 104.5 percent, or roughly 491,150 lbs. (2,185 kN) of thrust. Each SLS engine will operate at 109 percent thrust – approximately 512,300 lbs. (2,279 kN) maximum thrust in a vacuum.

The RS-25 uses a staged-combustion engine cycle that burns liquid hydrogen and liquid oxygen propellants at very high pressure. These engines operate in temperature extremes from -423 degrees Fahrenheit (-253 degrees Celsius) to 6,000 degrees Fahrenheit (3,316 degrees Celsius) and at pressures exceeding 7,000 psi (48,263 kPa). The engine was certified by ground testing during the shuttle program to operate at the higher thrust, though it was never used operationally.

With SLS, the RS-25 will experience several additional differences in performance requirements and operating environments. The SLS engine compartment will be colder because it is located directly below the liquid hydrogen fuel tank instead of the separate arrangement of the shuttle orbiter and external tank. The SLS engines will face higher liquid oxygen inlet pressures due to the higher position of the liquid oxygen tank relative to the engines as compared to the shuttle and external tank.

The Artemis I RS-25 liquid fuel engines operated successfully as part of the core stage Green Run hot fire test at Stennis in early 2021.

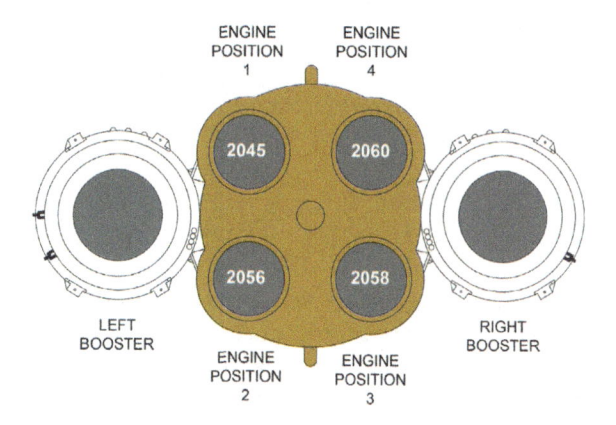

RS-25 ORIENTATION VIEW LOOKING UP FROM THE BOTTOM
Artemis engine serial numbers and their position in the core stage relative to the boosters.

**THE ELEMENTS OF SLS
RS-25 MANUFACTURING, TEST,
AND CHECKOUT**

In addition, the engine nozzles will also be subjected to a hotter launch pad environment due to their location closer to the SLS booster exhaust nozzles compared to a location farther above the booster exhaust nozzles on the space shuttle. Operationally, pre-launch engine conditioning will be different because of those environments, and the engine throttling and gimballing profile during ascent will be different due to vehicle acceleration and trajectory profile.

The SLS engines start in a staggered fashion: engine 1, engine 3, engine 4, then engine 2, approximately six seconds before booster ignition at T-0 and will follow this thrust profile:

– Engines start about 6 seconds prior to booster ignition.
– Engines reach 100 percent of rated power level about one second before booster ignition.
– Engines throttled to 109 percent rated power level at booster ignition (T-0).
– At T + 55 seconds, engines can be throttled down to lessen stress on the rocket during maximum dynamic pressure (max Q).
– At T + 81 seconds engines throttle back up to 109 percent rated power level.
– At T + 123 seconds engines throttle down to 85 percent rated power level (booster separation "bolt bucket" to reduce stress on attach struts and frangible bolts).
– At T + 132 seconds engines throttle back up to 109 percent rated power level following booster separation.
– At T + 421 seconds engines throttle back as needed to reduce acceleration forces (max g level).
– At T + 476 seconds engines throttle back to 67 percent rated power level.
– At T + 483 seconds, MECO.

The Artemis II RS-25 engines are designated as backups for the Artemis I engines. Engines 2047, 2059, 2062, and 2063 have completed modification and are ready for flight if needed.

Manufacturing, Test, and Checkout

Flight engines are processed by Aerojet Rocketdyne at NASA Stennis. The company also supports engine integration into the core stage at Michoud, engine testing at Stennis and vehicle integration and launch at Kennedy.

Because of differences in SLS requirements and operating environment, the SLS Program conducted a series of "adaptation" firing tests from 2015 to 2019 at Stennis with a pair of existing ground test RS-25s. Testing also included running the engines at 113 percent thrust to demonstrate new-production engines can run safely at 111 percent in case of a contingency.

The series also included Green Run testing of all 16 new engine controllers, as well as testing of new parts manufactured with new techniques aimed at restarting engine production. A total of 32 tests amassed nearly 15,000 seconds of hot fire time.

The 14 previously flown engines do not require certification firing tests because they are flight-proven and certified to operate at 109 percent thrust. The Stennis test stand was also used to Green Run test the two new engines built from shuttle components.

RS-25 engine and controller hot fire testing at Stennis.

Flight History of Artemis I RS-25 Engines

Engines installed on the Artemis I core stage are serial numbers 2045, 2056, 2058, and 2060. All were reconfigured for SLS and accepted for flight without additional hot fire testing, having flown previously. All meet the SLS engine life requirements of four starts and 1,700 seconds operation time remaining. All four RS-25s slated for Artemis I have a rich history in NASA's Space Shuttle Program flying to low-Earth orbit. They are veterans of these successful space shuttle missions:

Engine 2045: STS - 89, 95, 92, 102, 105, 110, 113, 121, 118, 127, 131, 135

- STS-89, Endeavour, flew in 1998, the eighth Shuttle-Mir docking.
- STS-95, Discovery, flew in 1998, Sen. John Glenn's flight.
- STS-92, Discovery, flew in 2000, ISS assembly flight, Zenith Z1 Truss, pressurized mating adapter (PMA-3), and was the 100th space shuttle flight.
- STS-102, Discovery, flew in 2001, ISS assembly flight, first use of Multi-Purpose Logistics Module (MPLM), Leonardo.
- STS-105, Discovery, flew in 2001, ISS assembly flight, second flight of Leonardo.
- STS-110, Atlantis, flew in 2002, ISS assembly flight, S0 Truss segment, Mobile Transporter.
- STS-113, Endeavour, flew in 2002, ISS assembly flight, P1 Truss.
- STS-121, Discovery, flew in 2006, ISS utilization mission, only launch on Independence Day, MPLM Leonardo.
- STS-118, Endeavour, flew in 2007, ISS assembly flight, starboard S5 truss, External Stowage Platform 3, final Spacehab logistics module flight.
- STS-127, Endeavour, flew in 2009, ISS assembly flight, JEM/Kibo Exposed Facility, record for most humans in space at the same time in same vehicle (13) and tied record of 13 people in space at one time (13).
- STS-131: Discovery, flew in 2010, ISS flight, MPLM.
- STS-135: Atlantis, flew in 2011, ISS utilization flight, MPLM, last shuttle mission.

Engine 2056: STS - 104, 109, 114, 121

- STS-104, Atlantis, flew in 2001, ISS assembly, installed Quest joint airlock module.
- STS-109, Columbia, flew in 2001, Hubble Space Telescope servicing mission.
- STS-114, Discovery, flew in 2005, ISS assembly flight, first mission since Columbia accident.
- STS-121, Discovery, flew in 2006, ISS logistics mission, Leonardo MPLM to Unity.

Engine 2058: STS - 116, 120, 124, 119, 129, 133

- STS-116, Discovery, flew 2006, ISS assembly flight, P5 truss.
- STS-120, Discovery, flew in 2007, ISS assembly flight, delivered Harmony module.
- STS-124, Discovery, flew in 2008, ISS assembly flight, delivered JEM/Kibo.
- STS-119, Discovery, flew in 2009, ISS assembly flight, starboard Integrated Truss Segment, solar arrays, batteries.
- STS-129, Atlantis, flew in 2009, ISS assembly flight, 2 ExPRESS logistics carriers.
- STS-133, Discovery, flew in 2011, ISS assembly flight, Permanent MPLM/Leonardo, 3rd ExPRESS carrier, Robonaut, last flight of Discovery.

Engine 2060: STS - 127, 131, 135

- STS-127, Endeavour, flew in 2009, ISS assembly flight, Kibo module.
- STS-131, Discovery, flew in 2010, ISS flight, MPLM.
- STS-135, Atlantis, flew in 2011, ISS utilization flight, MPLM, last shuttle mission.

THE ELEMENTS OF SLS
RS-25 FUN FACTS

RS-25 Fun Facts

- The thrust provided by the SLS RS-25 engines could keep eight Boeing 747s aloft.
- The RS-25 is so powerful that it could power 846,591 mi. of residential streetlights – a street long enough to go to the Moon and back, then circle the Earth 15 times.
- Four RS-25 engines push the SLS rocket 73 times faster than an Indianapolis 500 race car travels.
- The RS-25 could provide twice the power needed to move all 10 existing Nimitz-class aircraft carriers at 30 knots.
- The RS-25 generates about 20 percent more thrust at sea level than comparable kerosene engines using the same amount of hydrocarbon fuels.
- The RS-25 exhaust is clean, superheated water vapor.
- Each turbine blade powering the RS-25's high-pressure fuel turbopump produces more horsepower than a Corvette ZR1's 638 horsepower, and its airfoil is the size of a quarter.
- In the RS-25, coolant travels through the main combustion chamber in two milliseconds, increasing its temperature by 400 degrees Fahrenheit (204 degrees Celsius).
- Pressure within the RS-25 is equivalent to an ocean depth of three miles – about the same distance where Titanic lies below the surface of the Atlantic Ocean.
- Together, the SLS's four RS-25 engines gobble propellant at the rate of 1,500 gal. (5,678 L) per second during their eight minutes of operation – more than enough to drain an Olympic-size swimming pool in one second.
- Hot gases exit the RS-25's nozzle at 9,600 mph (15,450 km/h) – 13 times the speed of sound, or fast enough to go from Los Angeles to New York City in 15 minutes.
- While the RS-25 is about the same size and weight as the two turbojet engines on an F-15 fighter, it generates eight times more thrust, which it achieves in less than five seconds, operating between -400 degrees Fahrenheit (-240 degrees Celsius) and 6,000 degrees Fahrenheit (3,316 degrees Celsius).

Solid Rocket Boosters

A pair of solid rocket boosters attached to the core stage supply more than 75 percent of total SLS thrust for the first two minutes of flight, operating with the four RS-25 main engines. Based on the space shuttle solid rocket boosters, the SLS boosters are the largest, most powerful solid propellant boosters ever built for flight. Each SLS booster is 177 ft. (54.1 m) tall, 12 ft. (3.7 m) in diameter, and weighs 1.6 million lbs. (726 t) when filled with solid propellant.

The propellant on the SLS Block 1 rocket boosters consists of PBAN, ammonium perchlorate, and aluminum powder. Standing 17 stories tall and burning approximately 5.5 tons (5 t) of propellant each second, each booster generates more thrust than 14 four-engine jumbo commercial airliners.

The SLS booster is based on three decades of knowledge and experience gained with the space shuttle booster with several design, process, and testing improvements for greater performance, safety, and affordability. Shuttle-heritage hardware transferred to the SLS Program for early flights includes forward structures, metal cases, aft skirts, and thrust vector control elements.

Close-up view of the throat plug break up and expulsion from the nozzle during ignition of the Qualification Motor 2 full-scale test firing in Utah.

The major difference between the shuttle and SLS boosters is the addition of a fifth solid propellant segment to the four-segment shuttle booster, allowing SLS to send more weight to TLI than the shuttle lofted to low-Earth orbit. The larger SLS motor burns about three seconds longer and has more than 200,000 lbs. additional thrust. Other new design features of the SLS booster include:

– New manufacturing processes.
– New nozzle design three inches wider than the shuttle nozzle to support increased thrust.
– Modified propellant formulation with lowered burn rate.
– New grain design with 12-fin forward segment.
– New asbestos-free insulation and liner configuration.
– New avionics.
– Improved nondestructive evaluation processes.
– Aft booster attach rings moved 20 ft. (6 m) aft from the space shuttle external tank location to allow them to attach to the core stage engine section rather than the liquid hydrogen tank.

The SLS booster is optimized for a single use, to launch heavier payloads and reduce operational costs associated with the reusable booster used during the Space Shuttle Program. To that end, parachutes and other recovery features have been removed from the SLS booster. Deletion of the recovery systems translated into approximately 2,000 lbs. (0.9 t) additional payload to TLI, the maneuver that sends Orion to the Moon.

Each SLS solid rocket booster has three major assemblies: forward skirt, motor, and aft.

The forward skirt assembly includes the nose cap, frustum with four solid fuel booster separation motors, and forward skirt. The forward skirt houses booster avionics, the flight termination system, and the core stage attach support posts that carry most of the static and flight loads for the SLS stack.

THE ELEMENTS OF SLS
SOLID ROCKET BOOSTERS

The motor assembly has five segments, each with two case segments filled with rubbery propellant. Beginning from the aft, or rear, end of the booster, the segments are:

- Aft.
- Center aft.
- Center center.
- Center forward.
- Forward.

The forward segment contains the igniter that fires the length of the hollow motor segments to ignite all segments simultaneously. The segments, once ignited, burn from the inside out until all propellant is consumed.

The forward attach struts are also located on the forward segment. The aft motor segment is attached to an ablative exhaust nozzle that can be gimballed to steer the rocket. It contains the aft core stage attach struts.

The aft skirt assembly on each booster consists of the aft skirt, the thrust vector control system for the booster, and four solid booster separation motors. The thrust vector control system gimbals the exhaust nozzle and steers the vehicle. The boosters provide 90 percent of vehicle steering for the first two minutes of flight.

In addition to the booster motor, Northrop Grumman, lead booster contractor, also produces the booster separation motors used to push the boosters away from SLS at motor burnout. Each booster has four separation motors in the nose and four on the aft skirt for a total of 16 on the vehicle.

After the RS-25 engines start, booster ignition occurs at T-0 seconds and the vehicle lifts off the mobile launcher platform and begins its ascent. The propellant grain design inside the motor segments – the shape of the hole running the length of the motor – is designed to provide a maximum 3.6 million lbs. (16,014 kN) thrust for roughly 25 seconds, ramp down to about 2.8 million lbs. (12,455 kN) thrust as SLS passes through max Q, and then ramp up to about 3.3 million lbs. (14,679 kN) of thrust before beginning to tail off about 90 seconds into flight.

The forward assemblies for the SLS solid rocket boosters include aerodynamic nose cones, frustums, and forward skirts containing avionics.

The solid rocket boosters' aft assembly includes the aft skirts with thrust vector control system inside that helps steer SLS during ascent. The entire weight of the SLS rocket goes through the aft skirt structures.

THE ELEMENTS OF SLS
SOLID ROCKET BOOSTERS

Booster separation occurs at about 2 minutes 12 seconds, 142,000 ft. (43.3 km) at Mach 4.3. In order to separate from the vehicle, pyrotechnically activated separation bolts on the booster attach struts fire and four solid fuel separation motors at the top and bottom of each booster fire to push the empty boosters safely away from the core stage. Each booster separation motor generates about 20,000 lbs. (89 kN) of thrust. The boosters splash down in the Atlantic Ocean roughly 5.5 minutes after launch.

The five-segment SLS solid rocket boosters generate 7.2 million lbs. (32,027 kN) of thrust – more than 75 percent of the rocket's total thrust for the first two minutes of flight.

**THE ELEMENTS OF SLS
SOLID ROCKET BOOSTERS
MANUFACTURING, TEST, AND CHECKOUT**

Manufacturing, Test, and Checkout

SLS boosters are manufactured by Northrop Grumman in Utah and at Kennedy. In Utah, motor segment cases are lined with insulation and then filled with solid propellant. Each segment undergoes a rigorous nondestructive inspection process to confirm it is ready for flight. Motor segments are shipped by rail to Kennedy prior to flight.

Northrop Grumman builds the aft and forward skirt assemblies at several facilities near the launch site in Florida. Previously flown structures are refurbished at the Hangar AF complex at the Cape Canaveral Space Force Station, then moved to the Booster Fabrication Facility (BFF) at Kennedy. There, Northrop Grumman installs avionics, thrust vector control systems, booster separation motors, and thermal protection. These systems are thoroughly tested at the BFF before delivery to the EGS program along with the motor segments and aft exit cones delivered from Utah.

EGS then builds the overall aft assembly in the RPSF at Kennedy by attaching the aft motor segment to the aft skirt assembly and adding the lower core stage attach struts. That aft assembly is moved to the Vehicle Assembly Building and is the first element lowered onto the mobile launcher to begin the rocket assembly.

The rest of the motor segments are then stacked and pinned on top of each other: center aft, center center, center forward, and finally the forward motor segment. A forward assembly, consisting of a forward skirt and nose cone, is then attached to each booster. After booster stacking, the core stage is lowered between the boosters. Each booster is mated to the SLS core stage by braces on the forward and aft booster segments. On the mobile launcher, the booster aft skirts carry the entire load, or weight, of the SLS and Orion stack.

Artemis I booster motors travel by rail, left, from prime contractor Northrop Grumman facilities in Utah to Kennedy. Right, motor segments are offloaded, inspected, and stored at Kennedy's RPSF before transport to the Vehicle Assembly Building for stacking on the mobile launcher.

THE ELEMENTS OF SLS
SOLID ROCKET BOOSTERS
MANUFACTURING, TEST, AND CHECKOUT

The SLS five-segment solid rocket motor is the result of experience of the shuttle booster motor as well as extensive post-shuttle development. Northrop Grumman conducted numerous tests on subscale motors to evaluate specific components or materials such as the motor case insulation and nozzle materials and configuration. The company also conducted a series of full-scale development motor tests and qualification motor tests.

Three development motor tests in 2009, 2010, and 2011 were used to evaluate the five-segment design and various other design changes. The Development Motor-1 test was the first to evaluate thrust, roll control, acoustics, and motor vibrations using 650 data channels. The Development Motor-2 test motor was chilled to 40 degrees Fahrenheit (4.4 degrees Celsius) to verify performance of 53 test objectives using more than 760 data channels. Development Motor-3 was the most heavily instrumented motor to date, with more than 970 sensors collecting data for 37 test objectives after the motor was heated to 90 degrees Fahrenheit (32.2 degrees Celsius).

Subsequently, Northrop Grumman conducted two full-scale qualification motor tests at its Utah test facilities in 2015 and 2016. Test motors were again heated to 90 degrees Fahrenheit (32.2 degrees Celsius) and cooled to 40 degrees Fahrenheit (4.4 degrees Celsius), respectively. Engineers collected additional data on motor upgrades such as the new insulation and booster case liner, redesigned exhaust nozzle, etc. to support more than 100 test objectives. This is how solid rocket motor designs are determined to be ready for flight. Once these ground tests are complete, the manufacturing process is carefully controlled to ensure the flight boosters will perform as well as the ground test articles.

Boosters lead contractor Northrop Grumman and NASA hot-fired several full-scale solid rocket motors to help qualify them for the SLS Program.

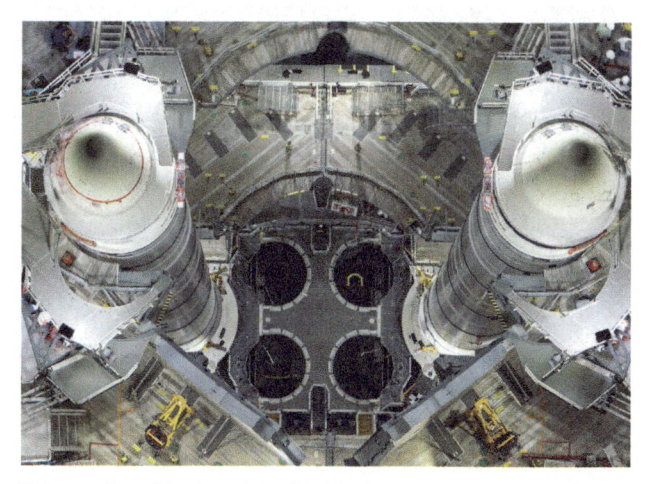

Birds-eye view of the Artemis I solid rocket boosters, the first element to be stacked on the mobile launcher in the Vehicle Assembly Building, awaiting the core stage.

THE ELEMENTS OF SLS
SOLID ROCKET BOOSTERS FLIGHT HISTORY

Flight History of the SLS Booster Structures

Several missions' worth of space shuttle booster hardware was transferred to the SLS Program for its early missions, while Northrop Grumman restarted booster manufacturing, incorporating several new features. The charts below summarize the shuttle flight history behind each of the Artemis I booster components.

Case Use History
SLS Artemis I-A (Left) Booster

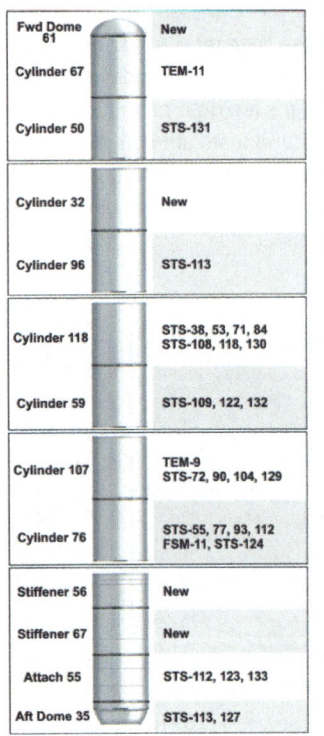

Component	History
Fwd Dome 61	New
Cylinder 67	TEM-11
Cylinder 50	STS-131
Cylinder 32	New
Cylinder 96	STS-113
Cylinder 118	STS-38, 53, 71, 84, STS-108, 118, 130
Cylinder 59	STS-109, 122, 132
Cylinder 107	TEM-9, STS-72, 90, 104, 129
Cylinder 76	STS-55, 77, 93, 112, FSM-11, STS-124
Stiffener 56	New
Stiffener 67	New
Attach 55	STS-112, 123, 133
Aft Dome 35	STS-113, 127

Component	History
Frustum 15	STS-41, 41G, 51I, 45, STS-55, 65, 69, 79, STS-86, 99, 108, 120, STS-131
Fwd Skirt 28	STS-53, 61, 67, 76, STS-84, 88, 92, 110, STS-115, 119
Aft Skirt 15	STS-41D, 51I, 46, 60, STS-73, 82, 91, 92, STS-107, 117, 128

Shuttle Flights
52

Static Tests
3

New
4

ETM – Engineering Test Motor
DM – Demonstration Motor
FSM – Flight Support Motor
FVM – Flight Verification Motor
STS – Space Transportation System
TEM – Technical Evaluation Motor

Case Use History
SLS Artemis I-B (Right)

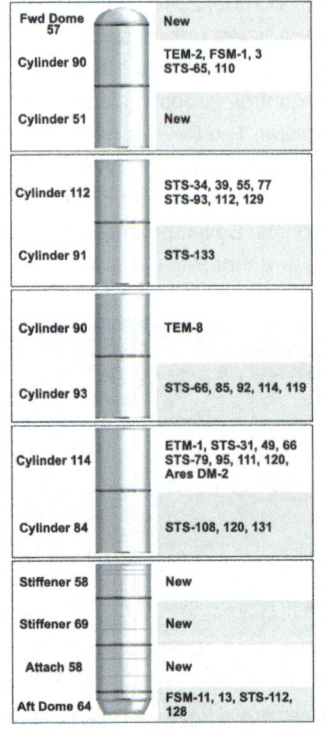

Component	History
Fwd Dome 57	New
Cylinder 90	TEM-2, FSM-1, 3, STS-65, 110
Cylinder 51	New
Cylinder 112	STS-34, 39, 55, 77, STS-93, 112, 129
Cylinder 91	STS-133
Cylinder 90	TEM-8
Cylinder 93	STS-66, 85, 92, 114, 119
Cylinder 114	ETM-1, STS-31, 49, 66, STS-79, 95, 111, 120, Ares DM-2
Cylinder 84	STS-108, 120, 131
Stiffener 58	New
Stiffener 69	New
Attach 58	New
Aft Dome 64	FSM-11, 13, STS-112, 128

Component	History
Frustum 14	STS-41G, 51I, 37, 49, STS-57, 64, 73, 94, STS-98, 113, 115, STS-126, 130
Fwd Skirt 27	STS-55, 66, 77, 85, STS-99, 105, 115, STS-119, 133
Aft Skirt 32	STS-27R, 32R, 43, STS-53, 68, 77, 89, STS-102, 119

Shuttle Flights
45

Static Tests
7

New
5

ETM – Engineering Test Motor
DM – Demonstration Motor
FSM – Flight Support Motor
FVM – Flight Verification Motor
STS – Space Transportation System
TEM – Technical Evaluation Motor

Artemis I booster aft assemblies ready for stacking in Kennedy Vehicle Assembly Building.

SLS Solid Rocket Boosters Fun Facts

- The SLS solid rocket boosters are 177 ft. (53.9 m) tall, taller than the Statue of Liberty from base to torch.
- Each booster is 12 ft. (3.6 m) in diameter.
- The boosters are the first element to be installed on the mobile launcher.
- The weight of the entire 5.75 million-lbs. (2,608 t) rocket rests on the booster aft skirts.
- Each booster weighs 1.6 million lbs. (726 t) when loaded with propellant.
- Each booster burns approximately 5.5 tons (5 t) of propellant a second.
- Each booster generates 3.6 million lbs. of thrust, more thrust than 14 four-engine jumbo commercial airliners.
- During the first two minutes of flight, the boosters provide more than 75 percent of the total SLS thrust.
- During hot fire testing in the Utah desert, the booster motor burns so hot that sand hit by the nozzle exhaust turns to glass.
- The boosters operate for about 2 minutes 12 seconds and are then jettisoned into the Atlantic Ocean.
- Once all the propellant in the solid rocket boosters has burned, 16 small solid rocket motors, called booster separation motors, in the forward and aft sections of each booster, fire simultaneously to safely push the boosters away from the SLS rocket.
- If their heat energy were converted to electric power, the two solid rocket boosters firing for two minutes would produce 2.3 million KWh of power, enough to supply power to more than 92,000 homes for a full day.

The completed Artemis I boosters are shown stacked on the mobile launcher in the Vehicle Assembly Building at Kennedy Space Center before integrating the other elements of SLS and Orion.

THE ELEMENTS OF SLS
INTEGRATED SPACECRAFT/PAYLOAD ELEMENT

Integrated Spacecraft/Payload Element

The SLS Block I configuration will be used for NASA's first three Artemis missions. The Integrated Spacecraft/Payload Element for the SLS Block I crew configuration is located above the core stage and below the Orion spacecraft. It includes elements for in-space propulsion, adapters, and a separation system.

On the Artemis I flight, the Integrated Spacecraft/Payload Element also includes a deployment system for secondary payloads. On Artemis I, the ICPS fires first to raise Orion's orbit and then a second time to send Orion on a path to the Moon. The launch vehicle stage adapter partially encloses the ICPS and serves as an interface between the larger core stage below and the smaller Orion stage adapter above. The Orion stage adapter connects the ICPS to Orion.

Jacobs crews bolt the launch vehicle stage adapter to the SLS core stage in the Vehicle Assembly Building.

Interim Cryogenic Propulsion Stage

The 45-ft. (13.7 m)-tall, 16.7-ft. (5.1 m)-diameter ICPS is a modified Delta Cryogenic Second Stage, a proven upper stage used on ULA's Delta IV family of launch vehicles. ULA builds the ICPS at its rocket factory in Decatur, Alabama. After completion, it's shipped to ULA's Delta Operations Center at Cape Canaveral Space Force Station on ULA's R/S Rocketship cargo ship for final checkout prior to stacking in the Vehicle Assembly Building.

Modifications to the Delta stage for the Block 1 SLS missions include lengthening the liquid hydrogen tank by 18.4 in. (46.7 cm), adding a second hydrazine bottle for attitude control, a new navigation system, Orion and launch vehicle stage adapter electrical and mechanical interfaces, a modified liquid hydrogen vent and relief valve, an RL10 in-flight helium injector purge to support engine restart, and RL10 qualification to SLS environments.

The stage is powered by liquid hydrogen and liquid oxygen, feeding a single Aerojet Rocketdyne RL10B-2 engine producing 24,750 lbs. (110 kN) of thrust. The RL10 has been in use more than 50 years to launch numerous military, government, and commercial satellites into orbit and send spacecraft to every planet in the solar system.

For the Artemis I mission, the ICPS will perform three burns:

- A perigee burn to circularize Orion's orbit.
- A TLI burn to push Orion out of Earth orbit to the Moon.
- A disposal burn after Orion separation to send ICPS into an orbit around the sun (heliocentric orbit).

The ICPS is a single-engine liquid hydrogen/liquid oxygen stage based on the Delta Cryogenic Second Stage from ULA that produces 24,750 lbs. (110 t) of thrust and performs the TLI burn to send Orion toward the Moon.

APPROXIMATE TIMELINE OF ICPS EVENTS DURING THE ARTEMIS I FLIGHT	
Major ICPS event	Approximate time post-launch (T-0)
ICPS/core stage separation	T-0 + 8 minutes and 30 seconds
ICPS coast to near apogee	T-0 + 8 minutes and 20 seconds to 51 minutes
ICPS perigee raise burn	T-0 + 50 minutes and 30 seconds to 51 minutes and 20 seconds
Parking orbit coast	T-0 + 51 minutes to 1 hour and 36 minutes
ICPS TLI burn	T-0 + 1 hour and 37 minutes to 1 hour and 55 minutes
Coast to Orion separation from ICPS	T-0 + 1 hour and 55 minutes to 2 hours and 5 minutes
Post-separation coast	T-0 + 2 hours and 5 minutes to 3 hours and 45 minutes
ICPS disposal maneuver to heliocentric orbit	T-0 + 3 hours and 45 minutes to 3 hours and 48 minutes

*Numbers based on 2021 mission analysis as this guide was in development and will vary based on season, launch date, and time. Numbers are approximated to meet security, export control, and proprietary requirements.

THE ELEMENTS OF SLS
LAUNCH VEHICLE STAGE ADAPTER

Launch Vehicle Stage Adapter

The cone-shaped launch vehicle stage adapter partially encloses the ICPS and provides the interface between the 27.6-ft. (8.4 m)-diameter core stage and 16.7-ft. (5.1 m)-diameter ICPS. The adapter is 27.5 ft. (8.4 m) tall. Hatches in the adapter allow technicians access to the ICPS during processing at Kennedy. The launch vehicle stage adapter also protects avionics and electrical devices in the ICPS from extreme vibration and acoustic conditions during launch and ascent.

A pneumatically actuated frangible joint assembly at the top of the launch vehicle stage adapter separates the ICPS, Orion stage adapter, and Orion from the core stage and launch vehicle stage adapter during ascent.

Cameras inside the adapter provide engineers with data on its performance and location in space after the core stage and launch vehicle stage adapter separate from the ICPS and Orion. The launch vehicle stage adapter stays attached to the core stage and they are disposed of together in the Pacific Ocean.

The launch vehicle stage adapter adapter partially encloses the ICPS and contains the pyrotechnic system to separate itself and the core stage from the ICPS and Orion.

Teledyne Brown Engineering manufactures the launch vehicle stage adapter at Marshall, where technicians manually apply spray-on insulation, which turns orange in sunlight.

THE ELEMENTS OF SLS
ORION STAGE ADAPTER

Orion Stage Adapter

The highest SLS element in the SLS stack, the Orion stage adapter, connects the ICPS to the Orion spacecraft. The Orion stage adapter is 18 ft. (5.5 m) in diameter, 5 ft. (1.5 m) tall, and is made of lightweight aluminum.

The adapter contains a diaphragm that provides a barrier to prevent gases, such as hydrogen, generated during launch from entering Orion. For Artemis I, the Orion stage adapter will also carry small secondary payloads, called CubeSats. SLS provided a comprehensive secondary payload deployment system for the CubeSats, including mounting brackets for commercial-off-the-shelf dispensers, cable harnesses, a vibration isolation system, and an avionics unit. Following separation from Orion, when the uncrewed spacecraft is a safe distance away, the avionics unit in the Orion stage adapter secondary payload deployment system will begin sending the signals to release the payloads at regular intervals (to ensure no re-contact) at pre-selected times. Following secondary payload deployment, the Orion stage adapter remains attached to the ICPS and enters a heliocentric disposal trajectory.

The completed Orion stage adapter after delivery from Marshall to Kennedy. The brackets to hold the dispensers for the CubeSat payloads and avionics unit are visible.

The ring-shaped Orion stage adapter is filled with 10 shoebox-sized CubeSats that will deploy when the ICPS separates from Orion on its way to the Moon.

57

THE ELEMENTS OF SLS
ISPE MANUFACTURING, TEST, AND CHECKOUT

Manufacturing, Test, and Checkout

The ICPS is manufactured by ULA. Its RL10 engine is manufactured by Aerojet Rocketdyne in its West Palm Beach, Florida facility.

Teledyne Brown Engineering manufactures the launch vehicle stage adapter using self-reacting friction-stir welding tools at the Advanced Weld Facility at Marshall. The adapter consists of two cones and two ring sections. Technicians first vertically weld panels to make the bottom cone. Then they weld panels to make the upper cone. Finally, a circumferential weld joins the two cones. Welded rings go on the top and bottom of the cone.

The first launch vehicle stage adapter manufactured at Marshall's Advanced Weld Facility was used as a test article along with the ICPS, Orion stage adapter, and a core stage simulator for a series of structural tests conducted at Marshall. Engineers subjected the stack of test articles to loads up to 40 percent greater than they will experience during ascent and flight. The results of this Integrated Structural Test were used to validate computer models simulating how the Integrated Spacecraft/Payload Element will behave in flight.

The Orion stage adapter is manufactured in-house at Marshall. Technicians in the Advanced Weld Facility have made three adapters in support of Artemis I. The first adapter was used in the Integrated Structural Test described above. It was later shipped to Lockheed Martin in Denver, where engineers used it for additional structural testing. The second adapter flew on Exploration Flight Test-1 in December 2014, connecting the Orion spacecraft to a ULA Delta IV launch vehicle for Orion's initial test flight. The third adapter is the Artemis I flight article.

The Integrated Structural Test series at Marshall was the beginning of the largest structural testing campaign since the Space Shuttle Program. The test qualified the launch vehicle stage adapter, ICPS, Orion stage adapter, and frangible joint assembly that separates the ICPS and Orion from SLS. The integrated stack, which included an Orion simulator, was subjected to forces up to 40 percent higher than anticipated flight loads. Approximately 50 test cases collected about 1,900 channels of data including temperature, deflection, torsion, compression, and other factors.

The Integrated Structural Test at Marshall included applying loads 40 percent greater than expected during launch to test versions of the launch vehicle stage adapter, frangible joint assembly, ICPS, Orion stage adapter, and a mass simulator for Orion.

The launch vehicle stage adapter structural test article for the Integrated Structural Test was manufactured and tested at Marshall.

SLS Avionics and Software

The SLS core stage, engines, boosters, and ICPS all have computers and software that monitor and control their functions. The avionics in the engines, boosters, and ICPS are connected to the flight avionics in the core stage. The core stage flight computers use data from the distributed avionics systems from the boosters, engines, and throughout the core stage to control the rocket and carry out its mission. It also transmits vehicle performance data to controllers on the ground on Artemis I and to the crew in Orion on later missions.

Core stage avionics and flight software serve as the "brains" of the rocket. They contain and execute the commands to prepare and launch SLS, route data and commands to the stage, distribute power, produce navigation and flight control data, produce range safety tracking data, execute flight termination commands, produce motion imagery, provide telemetry to ground systems, synchronize and process data, monitor stage conditions, and receive/execute flight safety commands.

Core stage avionics consist of four main subsystems:

- Flight control.
- Telemetry.
- Flight safety.
- Electrical power.

Avionics equipment is distributed among the forward skirt, intertank, and engine section in the core stage (the solid rocket boosters and ICPS also contain avionics).

Three flight computers and four power, data, telemetry, and navigation systems are located in the forward skirt. Each of the three flight computers uses three microprocessors. Each flight computer executes the same software for redundancy.

The intertank houses 26 avionics systems for power, power distribution, data receiving/handling, telemetry, and vehicle camera control. The engine section contains 10 avionics systems related to engine monitoring and vehicle navigation.

SLS flight software provides the flight and pre-flight software functions necessary for on-pad pre-launch procedures, launch, and ascent of SLS up to ICPS separation. The software was developed at Marshall.

The flight control system, led by three redundant flight computers, monitors the rocket's condition, senses vehicle motion, generates navigation and control data, actuates main propulsion system valves, monitors the main propulsion system and engine controls, and routes flight-critical commands to engine thrust vector control systems, and controllers. The flight computers have 256 MB of RAM each.

The core stage telemetry system includes radio and ethernet communications with the ground, telemetry control, engineering and development flight instrumentation, and a motion imagery system.

The flight safety system provides range tracking data and controls the rocket's self-destruct function located in the core stage and boosters.

The avionics power system distributes ground power, stores ground power for flight, and provides data to ground control centers.

Core stage avionics:
- Command and telemetry controller.
- Power distribution.
- Data acquisition.
- Camera equipment.
- Liquid level sensors.
- Rate gyro.
- Radio-frequency transmitter.

ICPS avionics

Launch vehicle stage adapter avionics:
- Two cameras for ICPS separation.

Core stage avionics:
- Flight computers.
- Command and telemetry controller.
- Inertial navigation equipment.
- Radio-frequency transmitter.

Booster avionics

Core stage avionics:
- Main propulsion valve control.
- Core stage thrust vector control.
- Rate gyro.

Core stage engine controllers

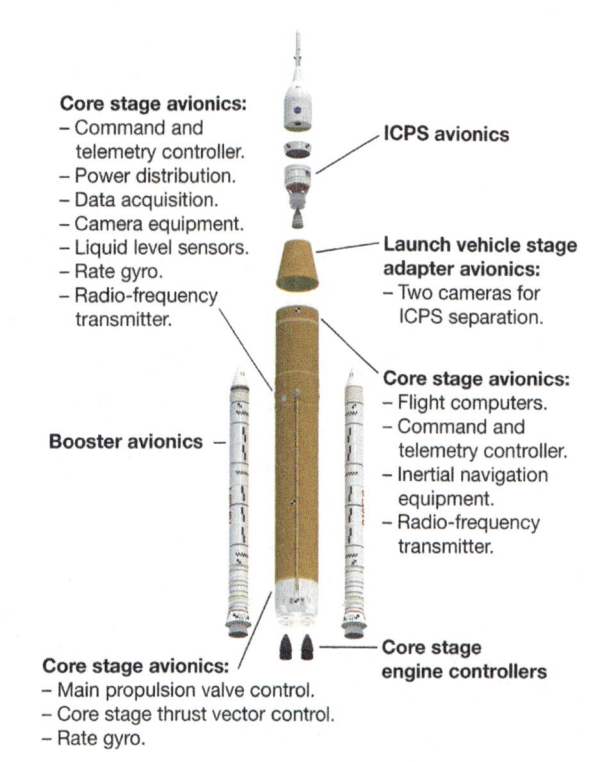

Avionics throughout the elements of SLS monitor the rocket's health and control the flight.

MANAGEMENT ROLES AND FACILITIES

MANAGEMENT ROLES AND FACILITIES

SLS is managed and supported by NASA Headquarters in Washington, D.C., and multiple NASA field centers across the nation. The Exploration Systems Development Mission Directorate and Space Operations Mission Directorate oversee NASA's human exploration programs in and beyond low-Earth orbit.

Marshall Space Flight Center

Marshall is home to the SLS Program Office and manages all areas of the program, including planning, procurement, development, testing, evaluation, production, and operation of the integrated vehicle. MSFC also developed and tested the flight software in-house.

The program office is also supported by Resident Management Offices (RMOs) at Stennis Space Center, which conducts engine and stage Green Run testing; and at Kennedy Space Center, responsible for integration and launch. The RMOs coordinate SLS technical and operations expertise with the EGS Program.

Unique Test Facilities

Several Marshall facilities support SLS. Building 4693 is a twin-tower open structure built to perform structural testing on a liquid hydrogen tank test article. Measuring 221 ft. (67.3 m) tall, the tower reuses the foundation of a former Saturn rocket test stand. For structural testing on the SLS liquid hydrogen tank structural test article, 38 hydraulic pistons were used to impart simulated flight loads on the tank, while more than 3,900 sensors measured temperature, deflection, strain, pressure, sound, and imagery.

Building 4697 is an L-shaped open tower structure built to test the liquid oxygen tank structural test article. Soaring 90 ft. (27.4 m) tall, the stand used 24 hydraulic pistons and more than 2,700 sensors to conduct 32 structural loads tests on the liquid oxygen tank structural test article.

Special test equipment consisting of two open-structure towers were built in Building 4619 at Marshall to test core stage engine section and intertank structural test articles.

The engine section tower measured 58 ft. (17.6 m) tall and the intertank tower measured 62 ft. (18.9 m) high. The engine section was subjected to 49 tests monitored by approximately 3,000 sensors. The intertank underwent 42 tests monitored by approximately 3,000 sensors.

The Systems Integration Lab/Systems Integration Test Facility allows engineers to test software with hardware in the loop. This allows them to fully simulate the integration of systems in virtual space prior to hardware manufacturing and test flight. Engineers in the lab also create and run end-to-end simulation environments in support of the entire project life cycle, including requirements development and analysis, as well as early prototyping, testing, and verification. These unique capabilities ensure that software and hardware integrate seamlessly before rocket manufacturing and assembly.

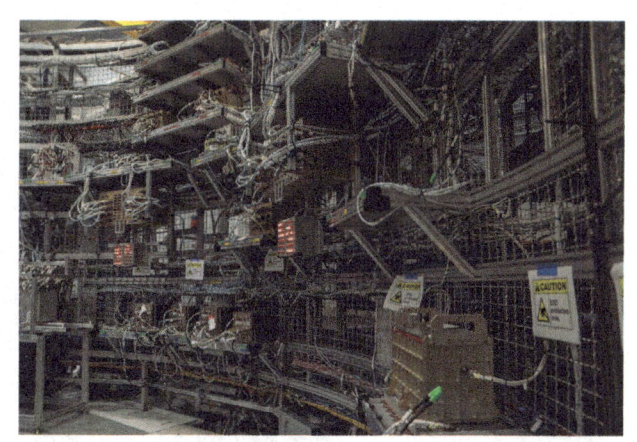

The Systems Integration Lab/Systems Integration Test Facility at Marshall is a full-scale mockup of flight-like avionics systems for the solid rocket boosters and core stage where engineers can test hardware and software integration prior to manufacture and flight.

**MANAGEMENT ROLES AND FACILITIES
MARSHALL SPACE FLIGHT CENTER**

Supporting Launch Operations

The SLS Engineering Support Center (SESC), located in the Huntsville Operations Support Center, allows engineers specializing in the engines, boosters, core stage, avionics, and upper stage to monitor the rocket's propulsion and other systems during the countdown and flight. The teams in the SESC leading up to and during launch analyze and monitor temperatures, pressures, flow rates, stresses, and other types of telemetry from the rocket. SESC teams also produce flash reports for the Mission Management Team and report on and archive data for additional study in the weeks and months after launch.

The SESC will support the LCC at Kennedy and the Flight Operations Directorate team located at the Mission Control Center at Johnson, performing in-depth analyses to help mission officials.

The SESC facility, which once served a similar purpose for both the Apollo and Space Shuttle Programs, has been updated with new communication and data equipment that provide greater data throughput and more voice communications channels. It also has the capability to reach out to SLS contractor organizations around the nation for additional expertise and attention to any issues. The SESC is staffed by a team from the SLS Program, MSFC engineering, and SLS contractors.

The Natural Environments Branch at Marshall characterizes terrestrial, space, and planetary natural environments in support of the SLS Program and other projects and programs across NASA. The Natural Environments Branch has been collecting detailed atmospheric data from the Earth's surface and aloft at Kennedy since the Apollo Program. The branch collects this data and develops weather databases for use by NASA and commercial customers to support spacecraft and launch vehicle programs in three major areas: system requirements design, verification and validation, and mission operations.

In addition to supporting SLS design and verification, the branch supports mission operations by generating complete atmospheric profiles for use in verifying SLS trajectory and loads constraints due to atmospheric wind conditions prior to launch.

For SLS and the Artemis I mission, the Natural Environments Branch also supports the Day of Launch Initialization Load Update. This update collects and uses day-of-launch atmospheric winds and temperature to design and verify the SLS flight profile in order to minimize loads (stresses) on the vehicle to ensure a safe flight.

On launch day, Kennedy and the Space Launch Delta 45 Eastern Range update weather data. The Marshall Natural Environments Branch uses the data to generate a profile of wind, temperature, density, and pressure at 100-ft. (30.5 m) intervals from the surface to 600,000 ft. (183 km). Johnson inputs that data into software to generate a flight guidance profile that is then validated by independent teams at Johnson and Marshall before sending those files to the launch team at Kennedy for upload onto SLS flight computers to translate into engine throttling and steering commands that minimize wind stresses on the rocket.

Marshall's SESC is networked to Kennedy, Johnson, and contractor locations nationwide to provide engineering support for SLS testing and launch operations.

**MANAGEMENT ROLES AND FACILITIES
MARSHALL SPACE FLIGHT CENTER**

On launch day, Marshall's Natural Environments Branch measures temperature, wind speeds, and more at Kennedy to provide data inputs critical to definiing final flight trajectories.

**MANAGEMENT ROLES AND FACILITIES
MICHOUD ASSEMBLY FACILITY**

Michoud Assembly Facility

The SLS core stage was manufactured, assembled, and checked out at Michoud before shipping to Stennis for stage Green Run testing and shipment to Kennedy for vehicle integration. For more than half a century, Michoud has been America's rocket factory, the nation's premier site for manufacturing, assembling, and checkout of large-scale space structures and systems.

The government-owned manufacturing facility is one of the largest in the world, with 43 acres of manufacturing space under one roof — a space large enough to contain more than 31 football fields. Marshall manages Michoud; commercial firms and NASA contractors use several areas of the facility. Michoud employs several key new manufacturing technologies and approaches to produce the SLS core stage:

- Lean manufacturing approaches with a production footprint about half what was used to manufacture the space shuttle's external tank. The SLS production workforce is also less than half of the external tank program headcount.
- Friction stir welding, providing stronger, lighter structures produced without welding defects. Core stage welded barrel sections used to assemble the rocket tanks are produced in less than half the time compared to space shuttle external tank production.
- Horizontal, single cell, robotically controlled application of spray-on foam insulation.
- Spun dome technology for dome caps, reducing complexity associated with gore panels.

Michoud is managed by Marshall and is the world's premier rocket factory and one of the largest manufacturing facilities in the world, with 43 acres under one roof.

The Vertical Assembly Center at Michoud is one of the world's largest welding tools and is used to execute vertical friction-stir welds of the SLS core stage liquid hydrogen tank.

**MANAGEMENT ROLES AND FACILITIES
MICHOUD ASSEMBLY FACILITY**

At Michoud, six major multi-function assembly/welding tools developed for SLS have resulted in a greater than 80 percent reduction in tooling from shuttle external tank production. This reduced tooling also minimizes hardware handling, reducing complex hardware lifting operations by more than 70 percent.
The new tools in NASA's rocket factory include:

- The Circumferential Dome Weld Tool, which performs circumferential friction-stir welds in the production of dome assemblies for the SLS core stage cryogenic tanks.
- The Gore Weld Tool performs vertical conventional friction stir welds in the production of gore assemblies for the SLS core stage tanks. (Gores are preformed aluminum alloy dome segments that are welded together to make the dome).
- The Circumferential Dome Weld and Gore Weld tools are special tooling for the Enhanced Robotic Weld Tool, which is used to make dome components for SLS.
- The Vertical Weld Center is a friction-stir-weld tool for wet and dry structures on the SLS core stage. It welds barrel panels together to produce whole barrels for the two pressurized tanks, the intertank, the forward skirt, and the engine section. It stands about three stories tall and weighs 150 tons (136 t).
- The Segmented Ring Tool uses a friction-stir-weld process to produce segmented support rings for the SLS core stage; the rings connect and provide stiffness between domes and barrels.
- The Vertical Assembly Center is where domes, rings, and barrels are joined to complete the tanks or dry structure assemblies. The tool also performs nondestructive evaluation on the completed welds. This tool measures 170 ft. (51.8 m) tall and 78 ft. (23.7 m) wide and is one of the world's largest welding tools.

The gore weld tool at Michoud is used to weld domes for the SLS core stage propellant tanks.

**MANAGEMENT ROLES AND FACILITIES
STENNIS SPACE CENTER**

Stennis Space Center

As the nation's largest propulsion test site, Stennis plays a major role in testing for SLS, just as it did for Saturn rockets and space shuttle vehicles.

The Fred Haise Test Stand was used to conduct multiple tests of RS-25 developmental engines to ensure the modified flight engines will operate to SLS requirements and environments. It was also used to Green Run test two new RS-25 engines, as well as new engine controllers for all 16 shuttle engines transferred from the Space Shuttle Program. The hot fire tests for the new engines and controllers also were conducted on Stennis's Fred Haise Test Stand. In 2021, NASA began using the stand to test RS-25 engines with newly designed parts made with advanced technology, such as additive manufacturing or 3D printing, for a new generation of engines for the fifth SLS flight and beyond.

Newer engines will have the same high performance as the upgraded engines used for early Artemis missions, but with anticipated manufacturing cost savings of more than 30 percent.

The Fred Haise Test Stand is a single-position, vertical-firing facility, which means that it can accommodate one rocket engine at a time and that engines are fired in an upright position with thrust directed downward. The stand was constructed from December 1964 to February 1967. The first test of an RS-25 rocket engine was conducted on the Fred Haise Test Stand on January 9, 2015. On April 4, 2019, Stennis completed testing of all 16 RS-25 main engines that will help launch the first four SLS missions.

NASA Stennis provides unique facilities for testing SLS RS-25 engines and controllers. The B Complex, foreground, was used for Green Run testing the SLS core stage.

**MANAGEMENT ROLES AND FACILITIES
STENNIS SPACE CENTER**

The stand extends 58 ft. (17.7 m) below ground and 158 ft. (48.2 m) above ground. It can withstand rocket engine thrust up to about 1.1 million lbs. (4,893 kN) of force; the thrust limit is known as the maximum dynamic load.

The B Test Complex at Stennis features a dual-position, vertical-firing test stand designated B-1/B-2, built in the 1960s. The B-1 Test Stand is designed for single-engine testing. The B-2 Test Stand is built to accommodate rocket stage testing. The B-2 Test Stand underwent major refurbishment and modifica¬tion to conduct the SLS Artemis I core stage Green Run test series to ready the stage for integration and flight.

The B-1/B-2 Test Stand is anchored in the ground with 144 ft. (43.9 m) of steel and concrete. As constructed, the soft core of the B-2 Test Stand was about 290 ft. (88.4 m) tall. The new steel superstructure added for testing SLS extends that height to almost 350 ft. (106.7 m), ranking the stand as one of the tallest structures in the state of Mississippi.

Stennis test stands are linked by a 7.5-mile canal system used for transporting rocket stages and liquid propellants. Support facilities for Stennis test stands include a test control center for each complex; data acquisition facilities; a large High Pressure Gas Facility to supply pressurized nitrogen, helium, hydrogen, and air; an electrical generation plant that provides power for engine tests to avoid potential disruptions in the power grid; and a High-Pressure Industrial Water Facility that features large diesel pumps, as well as a new electric pump, and a 66 million gal. (249,837,178 L) reservoir.

The main derrick crane atop the B-2 Test Stand was extended 50 ft. (15.2 m) with an increased load rating of 195 tons (177 t) to lift the SLS core stage, which is larger and heavier than the earlier Saturn V stages. More than 32,500 5/32-inch holes drilled in the B-2 Test Stand flame deflector direct more than 240,000 gal. (908,499 L) of water a minute to cool engine exhaust during a test. Another 92,000 gal. (348,258 L) of water per minute is sprayed through 92 nozzles to provide vibro-acoustic suppression protection to the core stage during testing. More than 100 water nozzles are arrayed across the test stand to provide a curtain of water over the length of the core stage and across the facility, if needed, to prevent damage in the event of a fire or cryogenic spill.

NASA Stennis provides unique facilities for testing SLS RS-25 engines and controllers.

INDUSTRY PARTNERS

INDUSTRY PARTNERS

Truly America's rocket, over the course of the SLS Program more than 1,100 companies from 48 states have contributed to designing, developing, manufacturing, testing, and supporting SLS, the rocket that will return NASA's human spaceflight program to the Moon. Several prime contractors head up the manufacturing, test, and assembly of the major elements of SLS, but they in turn are supported by hundreds of smaller suppliers across the U.S.

**INDUSTRY PARTNERS
AEROJET ROCKETDYNE**

Aerojet Rocketdyne

Aerojet Rocketdyne is the prime contractor for the four powerful RS-25 engines used to help propel each SLS mission. The four liquid hydrogen/liquid oxygen-fed RS-25 engines produce more than 2 million lbs. (8,896 kN) of thrust. The RS-25 contract is managed out of the company's Canoga Park, California, facility, which is also where the majority of the design work and component fabrication takes place.

Assembly and testing occur at the company's facility located at Stennis. In addition to the RS-25 engines, the company also provides propulsion that will be used throughout the Artemis I mission. Aerojet Rocketdyne designs, manufactures, and tests the RL10 engine that propels the ICPS in West Palm Beach, Florida, and a suite of propulsion for the Orion spacecraft is manufactured in Redmond, Washington. The company's subsidiary, ARDE, located in Carlstadt, New Jersey, builds the oxygen and nitrogen (nitrox) tank for the life support system on Orion and five composite overwrapped pressure vessels that store high-pressure helium to inflate Orion's flotation system upon water landing.

Industry partner Aerojet Rocketdyne, headquartered in Sacramento, California, supplies RS-25 engines for the SLS core stage and RL10 engines for the ICPS.

INDUSTRY PARTNERS
BOEING

Boeing

Boeing is the prime contractor for the design, development, test, and production of the SLS core stage and ICPS, as well as development of the flight avionics suite. Boeing built and tested the core stage for the Artemis I mission, and production for the Artemis II, III, and IV stages is underway, including the EUS, a more powerful in-space stage for NASA's Block 1B and ultimate Block 2 variants.

The Boeing SLS Program is managed out of the company's Space and Launch division in Huntsville, Alabama, and employs Boeing's workforce in Huntsville, at Michoud, Kennedy, and at other Boeing sites and with suppliers across the country. The Boeing Exploration Launch Systems office supports NASA on strategy and policy for Space Exploration programs procured by Marshall.

Boeing is the prime contractor for the SLS core stage, manufactured at Michoud, and the ICPS, manufactured by ULA.

INDUSTRY PARTNERS
NORTHROP GRUMMAN

Northrop Grumman

Northrop Grumman is the prime contractor for the design, development, testing, and production of the twin solid rocket boosters that provide more than 75 percent of initial thrust for SLS. Building off the knowledge and flight-proven hardware from the Space Shuttle Program, the new five-segment design features enhanced technologies and incorporates 25 percent more propellant to make these the largest rocket boosters ever built for flight. Additionally, Northrop Grumman provides 16 booster separation motors, designed to push the spent solid rocket boosters away from the core stage, for each launch.

The Northrop Grumman Northern Utah team manages the production and testing of the SLS solid rocket boosters, with teams at Kennedy and Huntsville, Alabama, overseeing various components and providing on-site support. Northrop Grumman also produces the launch abort motor and the attitude control motor for the Orion spacecraft's Launch Abort System. The abort motor is manufactured and tested out of the company's Promontory and Bacchus, Utah facilities, and work on the attitude control motor is based in Elkton, Maryland.

Northrop Grumman manufactures the solid rocket boosters for SLS, as well two of the three the solid motors used in Orion's launch abort system.

INDUSTRY PARTNERS
TELEDYNE BROWN ENGINEERING

Teledyne Brown Engineering

Teledyne Brown Engineering, based in Huntsville, Alabama, has been a leader in innovative products, systems, integration, operation, and manufacturing for the space industry for more than 65 years. Teledyne Brown provides engineering, technical support, and hardware for the launch vehicle stage adapter. The launch vehicle stage adapter is manufactured using the friction-stir welding tools in the Advanced Weld Facility at Marshall. It is the largest piece of the current SLS configuration to be built at Marshall. In addition to the Artemis I adapter, Teledyne Brown Engineering also delivered the launch vehicle stage adapter structural test article in 2016 for the Integrated Structural Test.

Teledyne Brown Engineering is the lead contractor for the launch vehicle stage adapter that has a pneumatically activated frangible joint that separates it and the core stage from the ICPS, Orion stage adapter, and Orion during flight.

INDUSTRY PARTNERS
UNITED LAUNCH ALLIANCE

United Launch Alliance

ULA is working collaboratively with Boeing to develop the 5-m ICPS for Artemis I, II, and III. The ICPS stages are manufactured in ULA's Decatur, Alabama, manufacturing facility. The ICPS stage is a modified version of the ULA 5-m Delta Cryogenic Second Stage, which has flown 24 times with 100 percent mission success.

With more than a century of combined heritage, ULA is the nation's most experienced and reliable launch service provider. ULA has successfully delivered more than 140 missions to orbit that aid meteorologists in tracking severe weather, unlock the mysteries of our solar system, provide critical capabilities for troops in the field, deliver cutting-edge commercial services, and enable GPS navigation.

ULA's program management, engineering, test, and mission support functions are headquartered in Denver, Colorado. Manufacturing, assembly, and integration operations are located at Decatur, Alabama, and Harlingen, Texas. Launch operations are located at Cape Canaveral Space Force Station, Florida, and Vandenberg Space Force Base, California.

ULA supplies the ICPS, which is based on the Delta Cryogenic Second Stage, for the SLS Block 1 rocket.

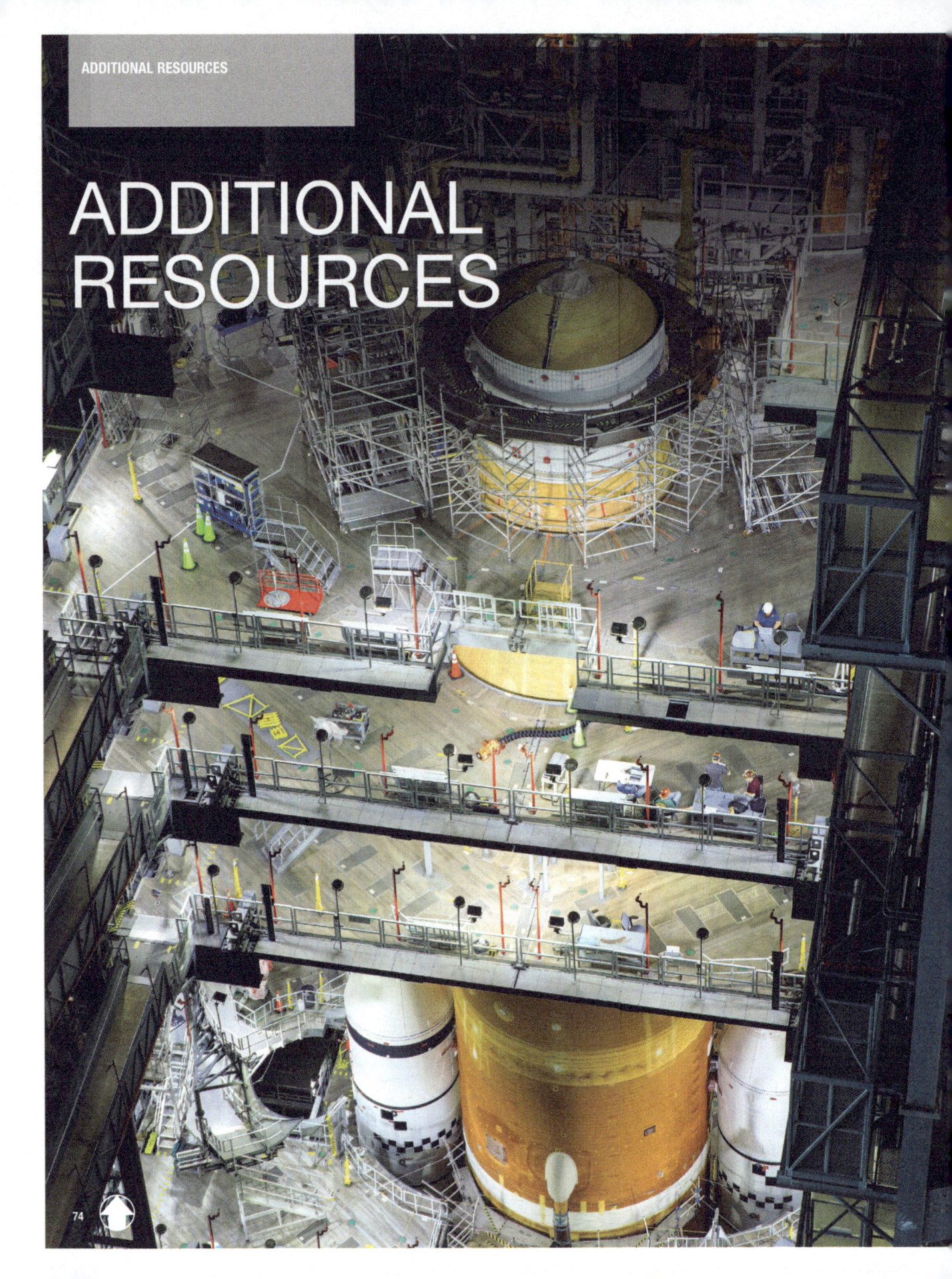

ADDITIONAL RESOURCES

ADDITIONAL RESOURCES

UNDERSTANDING SLS: INFOGRAPHICS

Understanding SLS: Infographics

In addition to the Quick Facts in the front of this document, selected SLS infographics below will help you understand SLS and its unique capabilities to enable continuous, ongoing human exploration of the Moon through the Artemis program.

Additional infographics describing SLS design, capabilities, testing, and more are available on the SLS web site at https://www.nasa.gov/exploration/systems/sls/multimedia/infographics.html

A BOLDER MISSION

Space Launch System

Space Launch System, or SLS, begins a bolder mission for NASA and the world — a new era of exploration unlike anything we've done before. Able to carry more payload than the space shuttle and generate more thrust at launch than the Saturn V, SLS will send the Orion spacecraft farther into space than Apollo ever ventured...and that's just the first flight!

Low-Earth Orbit:

Low-Earth orbit, or "LEO," is a term used to describe an orbit between 99 miles and 1,200 miles above the surface of the Earth. Relative to the size of the Earth, the green circle at right is approximately 250 miles thick and indicates the altitude at which the International Space Station currently orbits the Earth.

Apart from the lunar flights of the Apollo Program over 40 years ago, every human space flight in history has remained within the boundaries of LEO.

Exploring New Destinations:

Space Launch System is a deep space vehicle capable of extending our reach into the solar system. With unrivaled lift and unprecedented cargo capacity, NASA will launch people farther into space than previously possible.

www.nasa.gov/sls

UNDERSTANDING SLS: INFOGRAPHICS

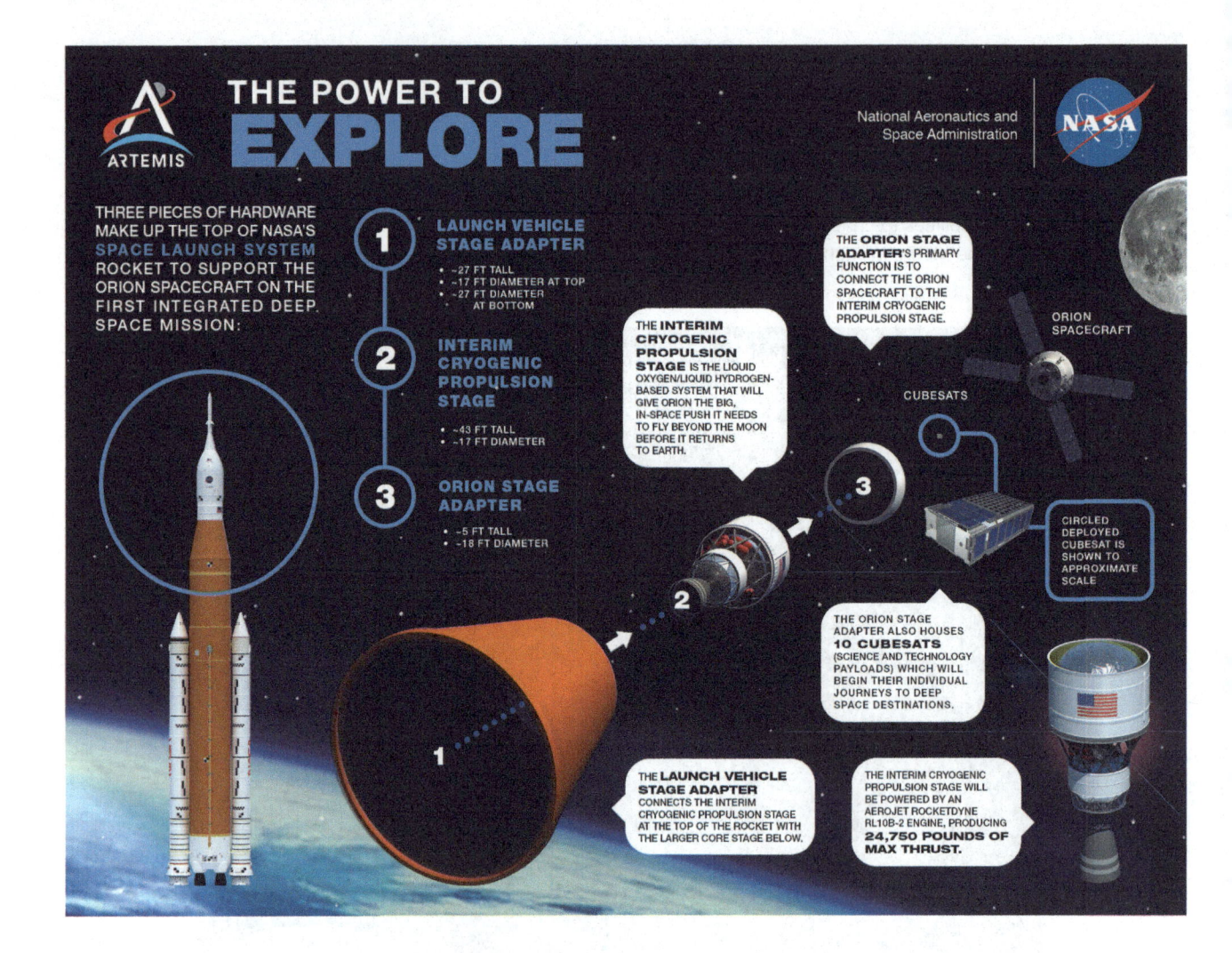

THE POWER TO EXPLORE

ARTEMIS

National Aeronautics and Space Administration — NASA

THREE PIECES OF HARDWARE MAKE UP THE TOP OF NASA'S **SPACE LAUNCH SYSTEM** ROCKET TO SUPPORT THE ORION SPACECRAFT ON THE FIRST INTEGRATED DEEP SPACE MISSION:

1. **LAUNCH VEHICLE STAGE ADAPTER**
   - ~27 FT TALL
   - ~17 FT DIAMETER AT TOP
   - ~27 FT DIAMETER AT BOTTOM

2. **INTERIM CRYOGENIC PROPULSION STAGE**
   - ~43 FT TALL
   - ~17 FT DIAMETER

3. **ORION STAGE ADAPTER**
   - ~5 FT TALL
   - ~18 FT DIAMETER

THE **INTERIM CRYOGENIC PROPULSION STAGE** IS THE LIQUID OXYGEN/LIQUID HYDROGEN-BASED SYSTEM THAT WILL GIVE ORION THE BIG, IN-SPACE PUSH IT NEEDS TO FLY BEYOND THE MOON BEFORE IT RETURNS TO EARTH.

THE **ORION STAGE ADAPTER**'S PRIMARY FUNCTION IS TO CONNECT THE ORION SPACECRAFT TO THE INTERIM CRYOGENIC PROPULSION STAGE.

ORION SPACECRAFT

CUBESATS

CIRCLED DEPLOYED CUBESAT IS SHOWN TO APPROXIMATE SCALE

THE ORION STAGE ADAPTER ALSO HOUSES **10 CUBESATS** (SCIENCE AND TECHNOLOGY PAYLOADS) WHICH WILL BEGIN THEIR INDIVIDUAL JOURNEYS TO DEEP SPACE DESTINATIONS.

THE **LAUNCH VEHICLE STAGE ADAPTER** CONNECTS THE INTERIM CRYOGENIC PROPULSION STAGE AT THE TOP OF THE ROCKET WITH THE LARGER CORE STAGE BELOW.

THE INTERIM CRYOGENIC PROPULSION STAGE WILL BE POWERED BY AN AEROJET ROCKETDYNE RL10B-2 ENGINE, PRODUCING **24,750 POUNDS OF MAX THRUST.**

SPACE LAUNCH SYSTEM CORE STAGE PATHFINDER
EXPERIENCE IS THE BEST TEACHER

Engineers are using the Core Stage Pathfinder, built in Cordova, Alabama, at multiple NASA centers.

The SLS Core Stage Pathfinder is a full-scale mockup made to be similar to the core stage in **SHAPE, SIZE AND WEIGHT** and was built to practice, practice, practice for upcoming activities with the largest rocket stage in the world—the SLS Core Stage.

The Pathfinder is made of steel and is the same size as the core stage

212' LONG
27.6' DIAMETER
WEIGHS 228,000 POUNDS

PATHFINDER PLACES

Pathfinder will validate ground support equipment, train handlers to transport the core stage on a variety of terrains with different vehicles, and show it can be integrated with facilities.

- CORDOVA
- STENNIS — BAY ST. LOUIS
- MICHOUD — NEW ORLEANS
- KENNEDY — CAPE CANAVERAL

- MICHOUD'S ROCKET FACTORY
- STENNIS' B-2 TEST STAND
- NASA'S BARGE PEGASUS
- KENNEDY'S VEHICLE ASSEMBLY BUILDING

UNDERSTANDING SLS: INFOGRAPHICS

UNDERSTANDING SLS: INFOGRAPHICS

SPACE LAUNCH SYSTEM
INSIDE THE SLS CORE STAGE
THE BACKBONE OF THE SLS ROCKET

ARTEMIS

HOW BIG IS THE CORE STAGE?
- **212'** tall and **27.6'** in diameter
- **~2.3M POUNDS** with propellant
- The largest rocket stage ever built
- Fuels the world's most powerful rocket

SLS reaches **MACH 23** (faster than **17,000 MPH**) in just **8.5 MINUTES**.

1 ENGINE SECTION
- Delivers propellants from the LH2 and LOX tanks to **4 RS-25 ENGINES**
- Avionics to steer engines
- Aft booster attach point

2 LIQUID HYDROGEN (LH2) TANK
- Holds **537,000 GALLONS** of liquid hydrogen cooled to **-423°F**

3 INTERTANK
- Joins **LH2** and **LOX** tanks
- Houses avionics and electronics
- Forward booster attach point

4 LIQUID OXYGEN (LOX) TANK
- Holds **196,000 GALLONS** of liquid oxygen cooled to **-297°F**

5 FORWARD SKIRT
- Houses flight computers, cameras, and avionics — the **"BRAINS"** of the rocket

Fuels 4 engines to produce a total **2 MILLION POUNDS** of thrust

733,000 GALLONS of propellant fill the **LH2** and **LOX** tanks together, enough to fill **63** large tanker trucks.

BIGGER TANKS. BOLDER MISSIONS.
www.nasa.gov/SLS

#ARTEMIS

UNDERSTANDING SLS: INFOGRAPHICS

SPACE LAUNCH SYSTEM
ARTEMIS I:
FOUR RS-25 ENGINES
UPGRADED & READY FOR A BOLD NEW MISSION

Shown here from the bottom, looking up at the rocket from below, four liquid hydrogen (**LH2**) and liquid oxygen (**LOX**)-fueled **RS-25 ENGINES** are arranged in a square pattern, like legs on a table, providing stability and even distribution of propulsion forces to the rocket. At launch, they will produce

2 MILLION POUNDS OF THRUST to help power the Space Launch System.

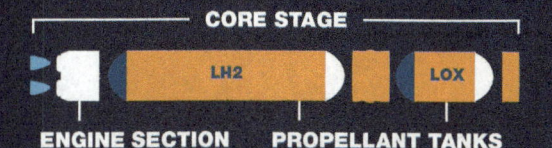

CORE STAGE
- ENGINE SECTION
- LH2
- LOX
- PROPELLANT TANKS

Including these first four engines, NASA has a total **16 ENGINES** available for the first four missions.

www.nasa.gov/SLS

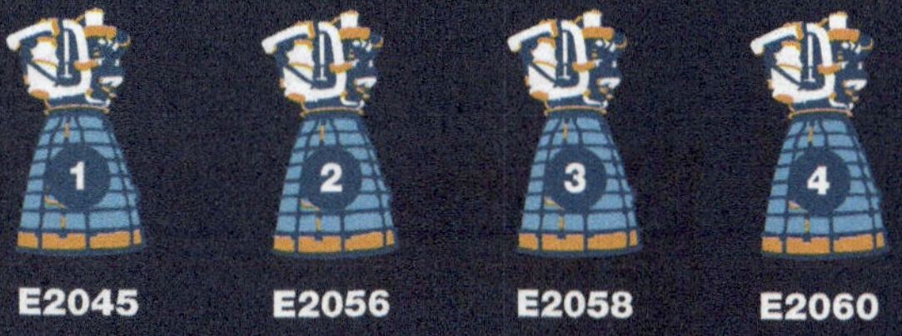

E2045   E2056   E2058   E2060

WHAT'S IN A NUMBER?
Each engine has its own number and NASA keeps a history of which engines are used on each mission. For the first SLS flight, they are engines **E2045, E2056, E2058** and **E2060**.

ENABLING MISSION SUCCESS
These four proven engines contributed to **21** successful shuttle flights over three decades.

- **E2045**: most veteran engine with **12 FLIGHTS** (First flight was STS-89 in January 1998 and last was STS-135 in July 2011); also flew on Astronaut John Glenn's last flight, STS-95
- **E2056**: Total **4 FLIGHTS**, including STS-114, NASA's Return to Flight after Columbia
- **E2058**: Total **6 FLIGHTS**, including first flight of a Swedish astronaut (Christer Fuglesang)
- **E2060**: Total **3 FLIGHTS**, including STS-135, the last shuttle mission

WHAT'S NEW FOR SLS?
- **ENGINE CONTROLLERS** - the brains of the engine
- **HIGHER THRUST** - equates with better performance
- **ADAPTED TO SLS UNIQUE OPERATING ENVIRONMENTS**

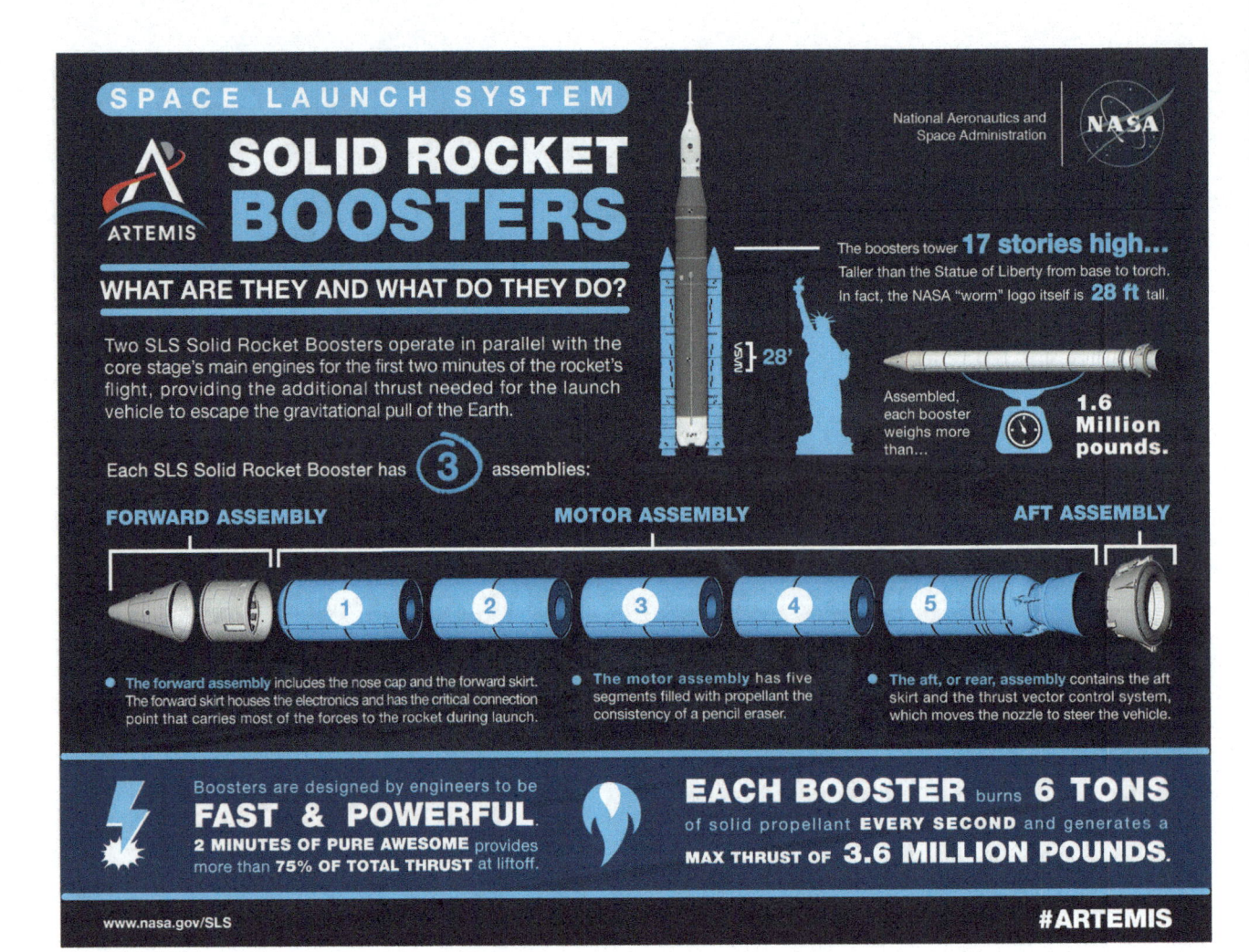

UNDERSTANDING SLS: INFOGRAPHICS

SPACE LAUNCH SYSTEM
SOLID ROCKET BOOSTERS

ARTEMIS

National Aeronautics and Space Administration — NASA

WHAT ARE THEY AND WHAT DO THEY DO?

Two SLS Solid Rocket Boosters operate in parallel with the core stage's main engines for the first two minutes of the rocket's flight, providing the additional thrust needed for the launch vehicle to escape the gravitational pull of the Earth.

The boosters tower **17 stories high…** Taller than the Statue of Liberty from base to torch. In fact, the NASA "worm" logo itself is **28 ft** tall.

Assembled, each booster weighs more than… **1.6 Million pounds.**

Each SLS Solid Rocket Booster has **3** assemblies:

FORWARD ASSEMBLY — **MOTOR ASSEMBLY** (1, 2, 3, 4, 5) — **AFT ASSEMBLY**

- **The forward assembly** includes the nose cap and the forward skirt. The forward skirt houses the electronics and has the critical connection point that carries most of the forces to the rocket during launch.
- **The motor assembly** has five segments filled with propellant the consistency of a pencil eraser.
- **The aft, or rear, assembly** contains the aft skirt and the thrust vector control system, which moves the nozzle to steer the vehicle.

Boosters are designed by engineers to be **FAST & POWERFUL**. 2 MINUTES OF PURE AWESOME provides more than **75% OF TOTAL THRUST** at liftoff.

EACH BOOSTER burns **6 TONS** of solid propellant **EVERY SECOND** and generates a **MAX THRUST** of **3.6 MILLION POUNDS.**

www.nasa.gov/SLS

#ARTEMIS

UNDERSTANDING SLS: INFOGRAPHICS

UNDER PRESSURE!

WHAT'S BEING TESTED AND HOW DO WE TEST IT?

THE **INTEGRATED STRUCTURAL TEST** IS SLS'S FIRST MAJOR STRUCTURAL TEST QUALIFYING **4 INTEGRATED TEST ARTICLES:**

1. LAUNCH VEHICLE STAGE ADAPTER
+ 2. FRANGIBLE JOINT ASSEMBLY
+ 3. INTERIM CRYOGENIC PROPULSION STAGE
+ 4. ORION STAGE ADAPTER

WHAT'S THE OBJECTIVE?

THIS TEST WILL VERIFY THAT EACH TEST ARTICLE CAN WITHSTAND THE EXPECTED FORCES OF FLIGHT.

THE TEST ARTICLES ARE FIRST STACKED JUST AS THEY WILL BE ON THE ROCKET, BETWEEN THE CORE STAGE AND THE ORION SPACECRAFT. THE TEST THEN SIMULATES THE TYPES OF PHYSICAL LOADS THIS **ENTIRE SECTION** OF THE ROCKET WOULD BE EXPECTED TO ENCOUNTER IN FLIGHT:

- COMPRESSION
- TENSION
- BENDING
- TORSION
- SHEAR

65 FT

3000+ BOLTS HOLD THE TEST ARTICLES AND SIMULATORS TOGETHER TO ENSURE THE VEHICLE IS SECURELY STACKED.

28 PISTONS APPLY PHYSICAL FORCE TO THE STACK:

FORCES AND PRESSURES ARE APPLIED TO TEST ARTICLES AT LOADS **40% ABOVE FLIGHT.**

100+ MILES OF CABLES TRANSMIT THE MEASUREMENTS FROM **1900 DATA CHANNELS.**

GOT FUEL?

INSTEAD OF FUEL, THE LIQUID HYDROGEN AND LIQUID OXYGEN FUEL TANKS ARE FILLED FOR TESTING WITH **NONFLAMMABLE LIQUID NITROGEN (LN2)*** AND THEN PRESSURIZED WITH GASEOUS NITROGEN TO SIMULATE FUELED FLIGHT CONDITIONS SAFELY.

* CRYOGENIC LIQUID NITROGEN AT **-320°F**

LN2

#ARTEMIS   www.nasa.gov/SLS

UNDERSTANDING SLS: INFOGRAPHICS

DESIGNED FOR DEEP SPACE
SPACE LAUNCH SYSTEM
THE ONLY ROCKET BUILT TO SEND MORE THAN 57,000 POUNDS TO DEEP SPACE

INTERIM CRYOGENIC PROPULSION STAGE (ICPS)
ONE RL10 ENGINE PROVIDES OVER 20,000 POUNDS OF THRUST TO SEND MORE THAN 26 METRIC TONS TO THE MOON THAT INCLUDES ORION, ASTRONAUTS AND SUPPLIES.

ORION SPACECRAFT
ORION WILL CARRY 4 ASTRONAUTS FROM EARTH TO THE MOON AND BRING THEM SAFELY HOME.

ORION STAGE ADAPTER
PROVIDES A PLACE TO CARRY SMALL SATELLITES TO DEEP SPACE WHERE THEY CONDUCT WORLD-CLASS SCIENCE FOR PENNIES ON THE DOLLAR.

LAUNCH VEHICLE STAGE ADAPTER
CONNECTS 27.6-FOOT DIAMETER CORE STAGE TO 16.5-FOOT DIAMETER ICPS AND PARTIALLY ENCLOSES THE ICPS IN-SPACE STAGE.

CORE STAGE
212-FOOT TALL STAGE HOLDS 733,000 GALLONS OF PROPELLANT TO POWER 4 RS-25 ENGINES FOR 8 MINUTES, SENDING THE ROCKET TO SPACE SOARING AT 17,000 MPH.

SOLID ROCKET BOOSTERS
EACH 17-STORY-TALL BOOSTER GENERATES 3.6 MILLION POUNDS OF THRUST, PROVIDING 75 PERCENT OF TOTAL THRUST DURING THE SLS ROCKET'S FIRST TWO MINUTES OF FLIGHT.

FOUR RS-25 ENGINES
MOST EFFICIENT ENGINES EVER BUILT, PROVIDING A TOTAL OF 2 MILLION POUNDS OF THRUST FOR LAUNCH AND ASCENT TO SPACE.

EARTH — OUTBOUND — DISTANCE TO MOON: 239,000 MI — RETURN — MOON — ORION

ACRONYM LIST

ACRONYM	DEFINITION
AES	Advanced Exploration Systems
BFF	Booster Fabrication Facility
EGS	Exploration Ground Systems
ESA	European Space Agency
EUS	Exploration Upper Stage
GSE	Ground Support Equipment
ICPS	Interim Cryogenic Propulsion Stage
ISS	International Space Station
ITCO	Intergrated Test and Check Out
JAXA	Japanese Aerospace Exploration Agency
LAS	Launch Abort System
LCC	Launch Control Center
LVSA	Launch Vehicle Stage Adapter
Max Q	Maximum Dynamic Pressure
MECO	Main Engine Cutoff
MPLM	Multi-Purpose Logistics Module
OSA	Orion Stage Adapter
PBAN	Polybutadiene Acrylonitrile
RPSF	Rotation, Processing, and Surge Facility
SESC	SLS Engineering Support Center
SLS	Space Launch System
SMD	Science Mission Directorate
SSME	Space Shuttle Main Engine
STS	Space Transporation System
TLI	Trans-Lunar Injection
ULA	United Launch Alliance

LINKS

For more information about SLS, visit:
http://www.nasa.gov/artemis
http://www.nasa.gov/sls
http://www.twitter.com/NASA_SLS
http://www.facebook.com/NASASLS
http://www.instagram.com/exploreNASA

NOTES

NOTES

National Aeronautics and Space Administration

George C. Marshall Space Flight Center
Huntsville, AL 35812
www.nasa.gov/marshall

www.nasa.gov

hsa llc

SLS-4071
NP-2022-08-65-MSFC

Made in the USA
Coppell, TX
06 December 2022